D1372714

DIOPHANTINE APPROXIMATIONS

Ivan Niven

DOVER PUBLICATIONS, INC.
Mineola, New York

Bibliographical Note

This Dover edition, first published in 2008, is an unabridged republication of the work originally published in 1963 by John Wiley & Sons, Inc., New York, as the Ninth Annual Series of Earle Raymond Hedrick Lectures of the Mathematical Association of America.

Library of Congress Cataloging-in-Publication Data

Niven, Ivan Morton, 1915–
    Diophantine approximations / Ivan Niven. — Dover ed.
        p. cm.
    Originally published: New York : Wiley, 1063, in series: Annual series of Earle Raymond Hedrick lectures of the Mathematical Association of America ; 9th.
    Includes bibliographical references and index.
    ISBN-13: 978-0-486-46267-7
    ISBN-10: 0-486-46267-6
    1. Diophantine analysis. I. Title.

QA242.N55 2008
512.7'3—dc22

                                                                2007046751

Manufactured in the United States of America
Dover Publications, Inc., 31 East 2nd Street, Mineola, N.Y. 11501

PREFACE

At the 1960 summer meeting of the Mathematical Association of America it was my privilege to deliver the Earle Raymond Hedrick lectures. This monograph is an extension of those lectures, many details having been added that were omitted or mentioned only briefly in the lectures. The monograph is self-contained. It does not offer a complete survey of the field. In fact the title should perhaps contain some circumscribing words to suggest the restricted nature of the contents, but such modifiers have been omitted for the sake of simplicity.

The topics covered are: basic results on homogeneous approximation of real numbers in Chapter 1; the analogue for complex numbers in Chapter 4; basic results on non-homogeneous approximation in the real case in Chapter 2; the analogue for complex numbers in Chapter 5; fundamental properties of the multiples of an irrational number, for both the fractional and integral parts, in Chapter 3. Many proofs are offered here for the first time, although the results themselves are not novel.

An attempt has been made, in a section entitled "Further results" at the end of each chapter, to provide a bibliographic account of closely related work. These sections also give the sources from which the proofs are drawn. Having used the literature freely, I wish to acknowledge especially the usefulness of the monographs by Cassels (1956) and Koksma (1936); see the bibliography for detailed references. The monographs by Cassels (1956) and Mahler (1961) treat many topics not considered here, such as the elegant work of Roth on the approximation of algebraic numbers. The topic of Diophantine approximations is also treated to some extent in the general texts on number theory, notably in Hardy and Wright (1960; Chapters 3, 10, 11, 23, 24) and LeVeque (1956; vol. 1, Chapter 9; vol. 2, Chapter 4).

A unique feature of this monograph is that continued fractions are not used. This is a gain in that no space need be given over to their description, but a loss in that certain refinements appear out of reach without the continued fraction approach. Another feature of this monograph is the inclusion of basic results in the complex case, which are often neglected in favor of the real number discussion. The parallel arguments for the real and complex cases in Chapters 2 and 5 are given here for the first time. This development of the theory was rounded out by the Theorem 5.3 of Eggan and Maier, who kindly provided me with their work prior to its publication.

Professor Herbert S. Zuckerman read the entire manuscript and made many helpful suggestions, for which I express my appreciation.   I also wish to thank Dr. Charles L. Vanden Eynden for suggesting certain clarifications.

IVAN NIVEN
EUGENE, OREGON

January 1963

CONTENTS

NOTATION AND CONVENTIONS

For any real number x:

$[x]$ denotes the greatest integer $\leq x$; that is, $[x]$ is the unique integer satisfying $[x] \leq x < [x] + 1$.

$\|x\|$ denotes the absolute value of the difference between x and the nearest integer; thus $\|x\| = \min |x-n|$, where the minimum is taken over all integers n.

(x) denotes the fractional part of x, namely $(x) = x - [x]$; this notation is used only in a few places where it is awkward to write $x - [x]$.

The symbol Z is used to denote the set of all integers.

$a|b$ means that the integer a is a divisor of the integer b.

$\{u_i\}$ denotes the sequence $u_1, u_2, u_3, \cdots$.

$\theta \equiv \lambda \pmod 1$ means that $\theta - \lambda$ is an integer.

N_α denotes the set of the integer parts of the multiples of α, thus

$$[\alpha], \quad [2\alpha], \quad [3\alpha], \quad \cdots.$$

The triangle inequality, $|u + v| \leq |u| + |v|$, is often used without reference or allusion.

A set $\theta_1, \theta_2, \cdots, \theta_n$ of real numbers is said to be linearly dependent over the rational numbers if there exist rational numbers $r_1, r_2, \cdots, r_n$ not all zero such that $\sum r_j \theta_j = 0$. Note that this is equivalent to saying that there are integers $k_1, k_2, \cdots, k_n$ not all zero such that $\sum k_j \theta_j = 0$. A finite set of real numbers is said to be linearly independent over the rational numbers in case they are not linearly dependent.

Wherever in the text a name appears with a date, such as Koksma (1935), the reference is to the paper or book of that date as listed in the bibliography at the end.

CHAPTER 1

The Approximation of Irrationals by Rationals

1.1. *The pigeon-hole principle*

Given a real number θ, how closely can it be approximated by rational numbers? To make the question more precise, for any given positive ε is there a rational number a/b within ε of θ, so that the inequality

$$\left| \theta - \frac{a}{b} \right| < \varepsilon$$

is satisfied? The answer is yes because the rational numbers are dense on the real line. In fact, this establishes that for any real number θ and any positive ε there are infinitely many rational numbers a/b satisfying the above inequality.

Another way of approaching this problem is to consider all rational numbers with a fixed denominator b, where b is a positive integer. The real number θ can be located between two such rational numbers, say

$$\frac{c}{b} \leqq \theta < \frac{c+1}{b},$$

and so we have $|\theta - c/b| < 1/b$. In fact, we can write

(1) $$\left| \theta - \frac{a}{b} \right| \leqq \frac{1}{2b}$$

by choosing $a = c$ or $a = c + 1$, whichever is appropriate. The inequality (1) would be strict, that is to say, equality would be excluded if θ were not only real but irrational. We shall confine our attention to irrational numbers θ because most of the questions about approximating rationals by rationals reduce to simple problems in linear Diophantine equations.

Now by use of the pigeon-hole principle (sometimes called the box principle) we can improve inequality (1) as in the following theorem. The pigeon-hole principle states that if $n + 1$ pigeons are in n holes, at least one hole will contain at least two pigeons.

THEOREM 1.1.   *Given any irrational number θ and any positive integer m, there is a positive integer $b \leqq m$ such that*

$$\|b\theta\| = |b\theta - a| < \frac{1}{m+1}.$$

The symbol a here denotes the integer nearest to $b\theta$, so that the equality $\|b\theta\| = |b\theta - a|$ holds by the definition of the symbolism.

Proof: Consider the $m + 2$ real numbers

(2) $0, 1, \theta - [\theta], 2\theta - [2\theta], \cdots, m\theta - [m\theta]$

lying in the closed unit interval.   Divide the unit interval into $m + 1$ subintervals of equal length

(3) $\dfrac{j}{m+1} \leqq x < \dfrac{j+1}{m+1}, \quad j = 0, 1, 2, \cdots, m.$

Since θ is irrational, each of the numbers (2) except 0 and 1 lies in the interior of exactly one of the intervals (3).   Hence two of the numbers (2) lie in one of the intervals (3); thus there are integers k_1, k_2, h_1, and h_2 such that

$$|(k_2\theta - h_2) - (k_1\theta - h_1)| < \frac{1}{m+1}.$$

We may presume that $m \geq k_2 > k_1 \geqq 0$.   Defining $b = k_2 - k_1$, $a = h_2 - h_1$, we have established the theorem.

Since $(m + 1)^{-1} < b^{-1}$, Theorem 1.1 implies that $\|b\theta\| < b^{-1}$. Furthermore, this inequality is satisfied by infinitely many positive integers b for the following reason.   Suppose there were only a finite number of such integers, say $b_1, b_2, \cdots, b_r$, with

$$\|b_j\theta\| < b_j^{-1} \quad \text{for } j = 1, 2, \cdots, r.$$

Then choose the integer m so large that

$$\frac{1}{m} < \|b_j\theta\|$$

holds for every $j = 1, 2, \cdots, r$. Then apply Theorem 1.1 with this value of m, and note that this process yields an integer b such that

$$\|b\theta\| < \frac{1}{m+1} < \frac{1}{m} < \|b_j\theta\|, \qquad j = 1, 2, \cdots, r.$$

Hence b is different from each of $b_1, b_2, \cdots, b_r$. Also $\|b\theta\| < b^{-1}$, so there can be no end to the integers satisfying this inequality. The following corollary states what we have just proved.

COROLLARY 1.2. *Given any irrational number θ, there are infinitely many rational numbers a/b, where a and $b > 0$ are integers, such that*

$$(4) \qquad\qquad \left|\theta - \frac{a}{b}\right| < \frac{1}{b^2}.$$

Note that this result is a considerable improvement over inequality (1). It is natural to ask whether Corollary 1.2 can also be improved, for instance, by the replacement of $1/b^2$ by $1/b^3$. It cannot; in fact, Corollary 1.2 becomes false if $1/b^2$ is replaced by $1/b^{2+\varepsilon}$ for any positive ε. Nevertheless, although the exponent cannot be improved, this corollary can be strengthened by a constant factor in (4). Specifically $1/b^2$ can be replaced by $1/(\sqrt{5}\,b^2)$, and no larger constant can be used than $\sqrt{5}$. This result, due to Hurwitz, is proved in the next section.

1.2. *The theorem of Hurwitz*

We first prove a preliminary result about Farey sequences. For any positive integer n, the Farey sequence F_n is the sequence, ordered in size, of all rational fractions a/b in lowest terms with $0 < b \leqq n$. For example,

$$F_7: \quad \cdots, \frac{-1}{6}, \frac{-1}{7}, \frac{0}{1}, \frac{1}{7}, \frac{1}{6}, \frac{1}{5}, \frac{1}{4}, \frac{2}{7}, \frac{1}{3}, \frac{2}{5}, \frac{3}{7}, \quad \cdots.$$

Of the many known properties of Farey sequences, only two are needed for our purposes, as follows.

THEOREM 1.3. *If a/b and c/d are two consecutive terms in F_n, then, presuming a/b to be the smaller, $bc - ad = 1$. Furthermore, if θ is any* given irrational number, and if r is any positive integer, then for all n sufficiently large the two fractions a/b and c/d adjacent to θ in F_n have denominators larger than r, that is, $b > r$ and $d > r$.

Proof: The proof of the first part is by induction on n. If $n = 1$, then $b = 1$, $d = 1$, and $c = a + 1$, so that

$$bc - ad = a + 1 - a = 1.$$

Next we suppose that the result holds for F_n, and prove it for F_{n+1}. Let a/b and c/d be adjacent fractions in F_n. First we note that $b + d \geqq n + 1$, since otherwise the fraction $(a + c)/(b + d)$, reduced if necessary, would belong to F_n. But this is not possible since

$$\frac{a}{b} < \frac{a + c}{b + d} < \frac{c}{d}.$$

Now with respect to F_{n+1} there are two possibilities: first that a/b and c/d are adjacent, and the second that some fraction or fractions lie between. In the first case there is nothing to prove because $bc - ad = 1$ by the induction hypothesis. In the second case, any such fraction, being in F_{n+1} but not in F_n, has denominator $n + 1$. Denoting the fraction by $k/(n + 1)$, we write

$$\frac{1}{bd} = \frac{c}{d} - \frac{a}{b} = \frac{c}{d} + \frac{k}{n + 1} + \frac{k}{n + 1} - \frac{a}{b}$$

$$= \frac{u}{d(n + 1)} + \frac{v}{b(n + 1)},$$

where $u = c(n + 1) - dk \geqq 1$, $v = bk - a(n + 1) \geqq 1$. Our aim is to establish that $u = 1$ and $v = 1$. If on the contrary $u > 1$ or $v > 1$ or both, then it follows that

$$\frac{1}{bd} > \frac{1}{d(n + 1)} + \frac{1}{b(n + 1)}, \qquad n + 1 > b + d,$$

which is contrary to what was established earlier. Hence $u = 1$ and $v = 1$, and so

$$\frac{c}{d} - \frac{k}{n + 1} = \frac{1}{d(n + 1)}, \qquad \frac{k}{n + 1} - \frac{a}{b} = \frac{1}{b(n + 1)}.$$

Finally, we observe that at most one fraction can occur between a/b and c/d in F_{n+1}. For if there were another fraction besides $k/(n+1)$ it must have the form $h/(n+1)$. Then the preceding argument implies that

$$\frac{h}{n+1} - \frac{a}{b} = \frac{1}{b(n+1)},$$

and hence

$$\frac{h}{n+1} - \frac{a}{b} = \frac{k}{n+1} - \frac{a}{b}, \qquad h = k.$$

This completes the proof of the first part of Theorem 1.3.

To prove the second part, let $m_1, m_2, \cdots, m_r$ denote the integers nearest to $\theta, 2\theta, \cdots, r\theta$. Choose n sufficiently large so that for every $j = 1, 2, \cdots, r$,

$$\frac{1}{n} < \left| \theta - \frac{m_j}{j} \right|.$$

If q is any integer, then for every $j = 1, 2, \cdots, r$,

$$|j\theta - m_j| \leq |j\theta - q|, \qquad \left| \theta - \frac{m_j}{j} \right| \leq \left| \theta - \frac{q}{j} \right|, \qquad \frac{1}{n} < \left| \theta - \frac{q}{j} \right|.$$

Now the difference between adjacent fractions in F_n does not exceed $1/n$, because F_n contains all fractions with denominator n, of which some are perhaps in reduced form. Hence if a/b and c/d are the fractions adjacent to θ in F_n, we see that

$$\left| \theta - \frac{a}{b} \right| < \left| \frac{c}{d} - \frac{a}{b} \right| \leq \frac{1}{n},$$

and

$$\left| \theta - \frac{c}{d} \right| < \left| \frac{c}{d} - \frac{a}{b} \right| \leq \frac{1}{n}.$$

A comparison of these with the previous inequalities establishes that $b > r$ and $d > r$, and the proof of Theorem 1.3 is complete.

Another result we shall need is the following.

LEMMA 1.4. *There are no positive integers x and y which satisfy simultaneously the inequalities*

$$(5) \quad \frac{1}{xy} \geq \frac{1}{\sqrt{5}} \left(\frac{1}{x^2} + \frac{1}{y^2} \right) \quad \text{and} \quad \frac{1}{x(x+y)} \geq \frac{1}{\sqrt{5}} \left(\frac{1}{x^2} + \frac{1}{(x+y)^2} \right).$$

Proof: If there were such integers, then from (5) it would follow that

$$0 \geqq x^2 + y^2 - \sqrt{5}xy \quad \text{and} \quad 0 \geqq (2 - \sqrt{5})(x^2 + xy) + y^2.$$

Adding these, we get

$$0 \geqq \tfrac{1}{2}\{(\sqrt{5} - 1)x - 2y\}^2$$

which is false for rational x/y.

We are now in a position to prove a basic result (Hurwitz, 1891).

THEOREM 1.5. *Given any irrational number θ there exist infinitely many rational numbers h/k in lowest terms such that*

$$(6) \qquad\qquad \left| \theta - \frac{h}{k} \right| < \frac{1}{\sqrt{5}k^2}.$$

Furthermore, this inequality is best possible in the sense that the result becomes false if $\sqrt{5}$ is replaced by any larger constant.

Proof: Locate θ between two consecutive fractions of the Farey sequence F_n, say $a/b < \theta < c/d$ with b and d positive. We consider two cases according to whether θ is greater or less than $(a + c)/(b + d)$. In case $\theta > (a + c)/(b + d)$, we prove that not all three of the inequalities

$$\theta - \frac{a}{b} \geqq \frac{1}{\sqrt{5}b^2}, \qquad \theta - \frac{a+c}{b+d} \geqq \frac{1}{\sqrt{5}(b+d)^2}, \qquad \frac{c}{d} - \theta \geqq \frac{1}{\sqrt{5}d^2}$$

can hold. For if we add the first and third of these, and then the second and third, we get (5) with $x = d$ and $y = b$.

In the other case, $\theta < (a + c)/(b + d)$, we prove that not all three of the inequalities

$$\theta - \frac{a}{b} \geqq \frac{1}{\sqrt{5}b^2}, \qquad \frac{a+c}{b+d} - \theta \geqq \frac{1}{\sqrt{5}(b+d)^2}, \qquad \frac{c}{d} - \theta \geqq \frac{1}{\sqrt{5}d^2}$$

can hold. For if we add the first and third of these, and then the first and second, we get (5) with $x = b$ and $y = d$.

Hence the inequality (6) holds with h/k replaced by at least one of a/b, c/d, and $(a + c)/(b + d)$. To prove that there are infinitely many solutions of (6) we argue as follows. Suppose there were only a finite number of solutions h/k, and we let r denote the maximum denominator among these solutions. Then the second part of Theorem 1.3 guarantees that for sufficiently large n the consecutive fractions a/b and c/d adjacent to θ in F_n have denominators greater than r. This process then gives a solution h/k to (6) of one of the three forms a/b, c/d, or $(a + c)/(b + d)$. Now a/b and c/d are in lowest terms by definition of Farey sequences. Also $(a + c)/b + d)$ is in lowest terms because

$$c(b + d) - d(a + c) = bc - ad = 1,$$

so that any common divisor of $a + c$ and $b + d$ is a divisor of 1. Thus the solution of (6) so obtained is in lowest terms and its denominator exceeds those of previously obtained solutions.

To complete the proof of the theorem we must show that $\sqrt{5}$ is the best possible constant. Before doing that, we remark on the proof thus far. The proof that (6) has infinitely many solutions has a slightly artificial aspect in that the inequalities (5) seem to have no motivating source. It might appear that some variation on the inequalities (5) would lead to better results. This is not the case, as we now show by establishing that the constant $\sqrt{5}$ in (6) is best possible.

Let θ_0 and θ_1 be defined by

$$\theta_0 = \frac{1 + \sqrt{5}}{2}, \qquad \theta_1 = \frac{1 - \sqrt{5}}{2},$$

so that

$$(x - \theta_0)(x - \theta_1) = x^2 - x - 1.$$

For any integers h and k, with $k > 0$, we see that

$$\left|\frac{h}{k} - \theta_0\right| \cdot \left|\frac{h}{k} - \theta_1\right| = \left|\left(\frac{h}{k}\right)^2 - \frac{h}{k} - 1\right| \neq 0.$$

Also $\theta_1 = \theta_0 - \sqrt{5}$ and so

$$\left|\frac{h}{k} - \theta_0\right| \cdot \left|\frac{h}{k} - \theta_1 + \sqrt{5}\right| = \frac{|h^2 - hk - k^2|}{k^2} \geq \frac{1}{k^2}.$$

An application of the triangle inequality gives

(7) $$\frac{1}{k^2} \leq \left|\frac{h}{k} - \theta_0\right| \cdot \left\{\left|\frac{h}{k} - \theta_0\right| + \sqrt{5}\right\}.$$

Now if for some positive number β there are infinitely many h_j/k_j, $j = 1, 2, 3, \cdots$, such that

(7a)
$$\left| \frac{h_j}{k_j} - \theta_0 \right| < \frac{1}{\beta k_j^2},$$

then $k_j \to \infty$ as $j \to \infty$. Furthermore, from (7) we get

$$\frac{1}{k_j^2} < \frac{1}{\beta k_j^2} \left(\frac{1}{\beta k_j^2} + \sqrt{5} \right),$$

$$\beta < \frac{1}{\beta k_j^2} + \sqrt{5},$$

$$\beta \leq \lim_{j \to \infty} \left(\frac{1}{\beta k_j^2} + \sqrt{5} \right) = \sqrt{5}.$$

Hence $\sqrt{5}$ is the largest possible constant in (6).

Thus the theorem is proved, and we note that the exponent 2 on the k^2 in (6) is best possible. That is, if γ is any fixed real number > 2, and c is any positive constant, there are only finitely many h/k satisfying

$$\left| \theta_0 - \frac{h}{k} \right| < \frac{1}{ck^\gamma}.$$

For if not, we could obtain infinitely many h/k satisfying (7a) with, say $\beta = 3$.

Next we formulate a simple consequence of Theorem 1.5.

COROLLARY 1.6. *Given any real numbers a_1, a_2, b_1, b_2 with $\Delta \neq 0$, where $\Delta = |a_1 b_2 - a_2 b_1|$, and given any positive ε, there are infinitely many pairs of integers h, k such that*

$$|a_1 k + b_1 h| \cdot |a_2 k + b_2 h| < \frac{\Delta}{\sqrt{5}} + \varepsilon.$$

Proof: If any one of a_1, a_2, b_1, b_2 is zero, or if a_1/b_1 or a_2/b_2 is rational, the result is immediate. So suppose that a_1/b_1 is irrational. Then let $-a_1/b_1$ play the role of θ in Theorem 1.5 and let h/k be any one of the rational numbers in that result. Define δ by

$$\frac{h}{k} + \frac{a_1}{b_1} = \frac{\delta}{k^2}$$

so that $|\delta| < 5^{-1/2}$. Thus we have $a_1 k + b_1 h = \delta b_1 / k$ and

$$
|a_1 k + b_1 h| \cdot |a_2 k + b_2 h|
$$

$$
= |\delta| \cdot \left| a_2 b_1 - a_1 b_2 + \frac{b_1 b_2 \delta}{k^2} \right|
$$

$$
\leqq |\delta| \cdot |a_2 b_1 - a_1 b_2| + \left| \frac{b_1 b_2 \delta^2}{k^2} \right|
$$

$$
< \frac{\Delta}{\sqrt{5}} + \left| \frac{b_1 b_2}{5 k^2} \right|.
$$

This proves the corollary, because $|b_1 b_2|/k^2$ can be made arbitrarily small by taking k sufficiently large.

1.3. *Asymmetric approximation*

The inequality of Theorem 1.5 can be written in the form

$$
- \frac{1}{\sqrt{5} k^2} < \theta - \frac{h}{k} < \frac{1}{\sqrt{5} k^2},
$$

so that the rational numbers h/k are allowable in symmetric fashion about the irrational θ. Next we prove an asymmetric result.

THEOREM 1.7. *Given any irrational number θ and any nonnegative real number τ, there exist infinitely many rational numbers h/k such that*

(8)
$$
- \frac{1}{(1 + 4\tau)^{1/2} k^2} < \theta - \frac{h}{k} < \frac{\tau}{(1 + 4\tau)^{1/2} k^2}.
$$

Furthermore, the statement holds if $\theta - h/k$ in (8) is replaced by $h/k - \theta$.

Note that Theorem 1.5 is a special case of this result, obtained by setting $\tau = 1$. We note also that the second sentence of Theorem 1.7 follows at once from the first; for if the first sentence is applied to the irrational number $-\theta$, the second part follows.

To prove Theorem 1.7 we begin with a preliminary result.

LEMMA 1.8. *Let θ be any irrational number, and τ any nonnegative real number. Let a/b and c/d be rational numbers with positive denominators such that $bc - ad = 1$ and*

$$(9) \qquad \frac{a}{b} < \frac{a + c}{b + d} < \theta < \frac{c}{d}.$$

Then (8) holds with h/k replaced by at least one of a/b, $(a + c)/(b + d)$, and c/d.

Proof: The argument is analogous to that in Theorem 1.5. Define λ and μ by

$$\lambda = (1 + 4\tau)^{-1/2} \quad \text{and} \quad \mu = \tau(1 + 4\tau)^{-1/2},$$

so that $\mu = (1 - \lambda^2)/4\lambda$ and $0 < \lambda \leqq 1$. Suppose that the lemma is false, so that

$$(10) \qquad \theta - \frac{a}{b} \geqq \frac{\mu}{b^2}, \qquad \theta - \frac{a + c}{b + d} \geqq \frac{\mu}{(b + d)^2}, \qquad \frac{c}{d} - \theta \geqq \frac{\lambda}{d^2}.$$

Adding the first and third of these inequalities and also the second and third, we conclude that

$$\frac{c}{d} - \frac{a}{b} = \frac{1}{bd} \geqq \frac{\mu}{b^2} + \frac{\lambda}{d^2}, \qquad \frac{c}{d} - \frac{a + c}{b + d} = \frac{1}{d(b + d)} \geqq \frac{\mu}{(b + d)^2} + \frac{\lambda}{d^2},$$

$$(11) \qquad \lambda b^2 - bd + \mu d^2 \leqq 0, \qquad \lambda(b + d)^2 - d(b + d) + \mu d^2 \leqq 0.$$

Adding these, we obtain

$$(12) \qquad 2\lambda b^2 + (2\lambda - 2)bd + (\lambda + 2\mu - 1)d^2 \leqq 0.$$

This quadratic form in b and d is seen to have discriminant zero since $\lambda^2 + 4\lambda\mu = 1$. Hence the quadratic form is a perfect square. In fact, if (12) is multiplied by the positive number 2λ, the result can be written

$$\{2\lambda b + (\lambda - 1)d\}^2 \leqq 0.$$

Equality must hold here since the square of a real number cannot be negative. It follows that equality must hold throughout all the relations (12), (11), and (10). Furthermore, we see that

$$2\lambda b + (\lambda - 1)d = 0, \qquad \lambda = \frac{d}{d + 2b},$$

so that λ is rational. But the third relation in (10) is now

$$\frac{c}{d} - \theta = \frac{\lambda}{d^2}.$$

This implies that θ is rational, which is a contradiction.

Thus Lemma 1.8 is proved, which we now use to prove Theorem 1.7.

First let a_1/b_1 and c_1/d_1 be the two consecutive fractions of the Farey series F_1 between which θ lies. Then $b_1 c_1 - a_1 d_1 = 1$ by Theorem 1.3. In case the inequalities

$$\frac{a_1}{b_1} < \frac{a_1 + c_1}{b_1 + d_1} < \theta < \frac{c_1}{d_1}$$

hold, then we apply Lemma 1.8 with a/b and c/d replaced by a_1/b_1 and c_1/d_1. Thus we would have one solution of (8). On the other hand, in case

$$\frac{a_1}{b_1} < \theta < \frac{a_1 + c_1}{b_1 + d_1} < \frac{c_1}{d_1},$$

let the positive integer j be chosen so that

(13) $$\frac{a_1}{b_1} < \frac{(j + 1)a_1 + c_1}{(j + 1)b_1 + d_1} < \theta < \frac{ja_1 + c_1}{jb_1 + d_1}.$$

This can be done because $(ja_1 + c_1)/(jb_1 + d_1)$ tends to a_1/b_1 as j increases indefinitely. Then we can apply Lemma 1.8 with a/b and c/d replaced by a_1/b_1 and $(ja_1 + c_1)/(jb_1 + d_1)$. The conditions (9) of the lemma are replaced by (13), and the condition $bc - ad = 1$ holds because

$$b_1(ja_1 + c_1) - a_1(jb_1 + d) = b_1 c_1 - a_1 d_1 = 1.$$

Thus in this case also we have one solution of (8). Let h_1/k_1 denote the solution of (8) obtained by use of F_1.

Next the second part of Theorem 1.3 is applied to choose n sufficiently large so that the two fractions of F_n adjacent to θ have denominators larger than k_1. Thus we get a_2/b_2 and c_2/d_2 in F_n with

$$\frac{a_2}{b_2} < \theta < \frac{c_2}{d_2}.$$

Then we repeat the preceding argument. That is, if

$$\frac{a_2}{b_2} < \frac{a_2 + c_2}{b_2 + d_2} < \theta < \frac{c_2}{d_2},$$

we apply Lemma 1.8 directly. Otherwise we choose j so that

$$\frac{a_2}{b_2} < \frac{(j + 1)a_2 + c_2}{(j + 1)b_2 + d_2} < \theta < \frac{ja_2 + c_2}{jb_2 + d_2}.$$

In one case or the other we get a solution h_2/k_2 of (8) with k_2 equal to one of

$$b_2, \ d_2, \ b_2 + d_2, \ jb_2 + d_2, \ (j + 1)b_2 + d_2.$$

It may be noted that all the fractions in the preceding inequalities are in lowest terms. For example,

$$b_2(ja_2 + c_2) - a_2(jb_2 + d_2) = b_2c_2 - a_2d_2 = 1,$$

so that the greatest common divisor of $ja_2 + c_2$ and $jb_2 + d_2$ is 1. Hence the solution h_2/k_2 of (8) is different from the first solution h_1/k_1.

Since this process can be repeated indefinitely, giving a new solution of (8) each time, the theorem is proved.

It might appear from an examination of the proof that a stronger or different result might be obtained by some refinement of the procedure leading from (11) to (12). That is, instead of simply adding the inequalities (11), what would happen if we used weighting factors in the addition process? It is not difficult, however, to prove that nothing new can be obtained by such a procedure.

The following consequence of Theorem 1.8 is obtained by taking $\tau = 0$.

COROLLARY 1.9. *Given any irrational θ there are infinitely many rationals h/k such that*

$$-\frac{1}{k^2} < \theta - \frac{h}{k} < 0,$$

and also infinitely many rationals h/k such that

$$0 < \theta - \frac{h}{k} < \frac{1}{k^2}.$$

1.4]   FURTHER RESULTS   13

1.4. *Further results*

The method of Section 1.1 can be extended to the problem of simultaneous approximation of several irrational numbers and the following
result can thus be obtained at once. Given any n real numbers
$\theta_1, \theta_2, \cdots, \theta_n$, there exist infinitely many sets of integers $(a_1, a_2, \cdots, a_n, b)$ with b positive such that

$$\left| \theta_j - \frac{a_j}{b} \right| < \frac{1}{b^{1+(1/n)}},$$

for all $j = 1, 2, 3, \cdots, n$. But even for $n = 2$ there is no analogue to
Theorem 1.5 (Hurwitz, 1891) giving the best possible constant in the
inequalities. For a general statement on this topic, see Davenport
(1954a), and also Davenport (1954b).

The proof of Theorem 1.5 in Section 1.2 was suggested by work of
Khintchine (1935) and LeVeque (1953). Theorem 1.7 in Section 1.3
is due to Segre (1945); the proof given here is by Niven (1962).

As an example of another result on asymmetric approximation we
cite the following theorem of Robinson (1947): Given any irrational
θ and any positive ε there are infinitely many rationals a/b such that

$$-\frac{1}{(\sqrt{5} - \varepsilon)b^2} < \theta - \frac{a}{b} < \frac{1}{(\sqrt{5} + 1)b^2}.$$

Corollary 1.9, a standard result in the continued fraction approach to
irrational numbers, implies Corollary 1.2, but not conversely. Eggan
and Niven (1961) have pointed out that Corollary 1.9 is best possible
in the following sense: Given any $\gamma > 1$ there are irrational numbers θ
such that

$$0 < \theta - \frac{a}{b} < \frac{1}{\gamma b^2}$$

has no solutions in rational numbers a/b; a similar result holds for
approximation with $\theta - a/b < 0$. For other work on asymmetric or
unsymmetric approximation see Olds (1946), Negoescu (1948),
Robinson (1948), Tornheim (1955a), and Eggan (1961).

Another variation on Theorem 1.5 arises if the rational numbers
a/b are restricted in some way, for example, if a and b must both be odd

or if a and b must satisfy some more general congruence conditions. Results along these lines have been obtained by Scott (1940), Robinson (1940), Oppenheim (1941), Hartman (1949), Koksma (1951), Tornheim (1955b), and Eggan (1961).

Theorem 1.5 is an assertion about every irrational number θ. If the irrational number $(\sqrt{5} - 1)/2$ and all numbers equivalent to it (in a sense explained in the next sentence) are deleted, then to each remaining irrational θ there are infinitely many rational numbers h/k satisfying

$$\left| \theta - \frac{h}{k} \right| < \frac{1}{\sqrt{8}\,k^2}.$$

Two real numbers θ and θ' are said to be *equivalent* if there exist integers p, q, r, s such that

$$\theta' = \frac{p\theta + q}{r\theta + s} \qquad \text{with } ps - qr = \pm 1.$$

Note that this is a so-called equivalence relation, and that the set of numbers equivalent to a fixed θ is countable. If next the number $\sqrt{2}$ and all equivalent numbers are also deleted from the set of irrational numbers, then to each remaining irrational number θ there are infinitely many rational numbers h/k satisfying

$$\left| \theta - \frac{h}{k} \right| < \frac{1}{b^2\sqrt{221}/5}.$$

This process can be continued, giving the Markoff chain of constants $\sqrt{5}$, $\sqrt{8}$, $\sqrt{221}/5$, $\sqrt{1517}/13$, $\cdots$, with limit 3. This theory is given in full detail in Cassels (1957, Chapter 2).

The Hurwitz Theorem 1.5 states that to each irrational θ there correspond infinitely many rationals h/k satisfying a certain inequality. Can the inequality be improved if only one rational is required? Yes. Prasad (1948) has determined, for each positive integer n, the best possible constant c_n such that to every irrational θ there correspond at least n rational numbers h/k satisfying

$$\left| \theta - \frac{h}{k} \right| < \frac{1}{c_n k^2}.$$

Eggan (1961) has extended this work by imposing on θ the kinds of restrictions mentioned in the preceding paragraph.

The last part of the proof of Theorem 1.5, in which it is shown that $\sqrt{5}$ is the best possible constant, is a special case of a method of Liouville, who proved that if θ is a real algebraic number of degree $n \geq 2$ and if there are infinitely many rationals h/k satisfying

$$\left| \theta - \frac{h}{k} \right| < \frac{1}{k^\beta}$$

for some real number β, then $\beta \leq n$. This result was improved several times, notably by Thue, Siegel, and Dyson. Finally, Roth (1955) proved the best possible result, which had been conjectured for some time, that $\beta \leq 2$. An estimate of the number of solutions of the inequality with $\beta = 2 + \varepsilon$ has been given by Davenport and Roth (1955). Ridout (1957) has formulated an extension of Roth's theorem wherein the class of rational approximations is restricted. An exposition of Roth's theorem is given in Cassels' (1957) monograph; an extension to the approximation of algebraic numbers by algebraic numbers can be found in LeVeque (1956, vol. II, Chapter 4). Mahler (1957) has given an application of Roth's theorem, including a sharpening of what is known about the Waring problem. Roth (1960) has outlined possible extensions and limitations of the method used in the proof of his theorem.

Corollary 1.6 holds (for at least one pair of integers h, k) with the ε removed but without a strict inequality; see Macbeath (1947) or Hardy and Wright (1960, p. 401).

CHAPTER 2

The Product of Linear Forms

The discussions of Chapter 1 can be looked at thus: Given any irrational number θ, we were concerned with integer values of k such that $k\theta$ would be close to an integer. Or, stated otherwise, we were asking for values of k such that $\|k\theta\|$ is small. It was established that there are infinitely many positive integers k so that $k\|k\theta\| < 5^{-1/2}$. Suppose now that we seek values of k such that $k\theta$ is close to an integer plus $\frac{1}{2}$. Then we would want $\|k\theta + \frac{1}{2}\|$ to be small. In general, for any real number α what can we say about $\|k\theta + \alpha\|$? Given any irrational θ and any real α, can we find infinitely many k such that $k\|k\theta + \alpha\| < 5^{-1/2}$? Yes, and moreover the constant can be improved from $5^{-1/2}$ to $\frac{1}{4}$ in the strictly nonhomogeneous case; for details see Corollary 2.4.

The expression $k\|k\theta + \alpha\|$ is the same as $k|k\theta + h + \alpha|$ for the appropriate integer h, and this is merely the special case of a product of two linear forms

$$|a_1 k + b_1 h + c_1| \cdot |a_2 k + b_2 h + c_2|,$$

where $a_1 = 1$, $b_1 = 0$, $c_1 = 0$, $a_2 = \theta$, $b_2 = 1$, $c_2 = \alpha$. This suggests the question: Can we choose integers h and k to make this product of two linear forms small? How small? These questions are also answered in this chapter.

2.1. *The Minkowski results*

We begin with a couple of preliminary results.

LEMMA 2.1. *Given any integers h, k such that $(h, k) = 1$ and given any real number ρ, there exist integers r and s such that $|ks - hr + \rho| \leq \frac{1}{2}$.*

Proof: Since h and k are relatively prime, the set $\{ks - hr\}$ runs through all integers as s and r run through all integers. Hence s and r can be chosen so that $ks - hr$ is the integer nearest to $-\rho$, and thus the lemma is proved.

LEMMA 2.2. *If α and β are any real numbers, then there is an integer u such that $|\alpha - u| < 1$ and*

(1) $$|\alpha - u| \cdot |\beta - u| \leq \tfrac{1}{4}$$

or

(2) $$|\alpha - u| \cdot |\beta - u| < \tfrac{1}{2}|\alpha - \beta|.$$

Proof: If α is an integer, set $u = \alpha$; otherwise we define the integer n by $n < \alpha < n + 1$. We treat two cases separately, according to whether β satisfies $n \leq \beta \leq n + 1$ or not. If β satisfies these inequalities, then $|n - \beta\| n + 1 - \beta| \leq \tfrac{1}{4}$; similarly for α, and so

$$|n - \alpha| \cdot |n - \beta| \cdot |n + 1 - \alpha| \cdot |n + 1 - \beta| \leq \tfrac{1}{16}.$$

Hence (1) holds with $u = n$ or $u = n + 1$.

There remain the possibilities $n > \beta$ and $\beta > n + 1$; these are symmetric, and we treat the case $n > \beta$. Using the standard inequality $2\sqrt{cd} \leq c + d$ for nonnegative real numbers c and d, we note that

$$2(n - \beta)^{1/2}(n + 1 - \beta)^{1/2}(\alpha - n)^{1/2}(n + 1 - \alpha)^{1/2}$$
$$\leq (n - \beta)(n + 1 - \alpha) + (\alpha - n)(n + 1 - \beta) = \alpha - \beta.$$

Squaring, and inserting absolute value signs, we get

$$|\alpha - n| \cdot |\beta - n| \cdot |\alpha - n - 1| \cdot |\beta - n - 1| \leq \tfrac{1}{4}|\alpha - \beta|^2,$$

and so we conclude that

$$|\alpha - u| \cdot |\beta - u| \leq \tfrac{1}{2}|\alpha - \beta|$$

with $u = n$ or $u = n + 1$. To establish (2), we must rule out the possibility of equality. Suppose then that

$$|\alpha - u| \cdot |\beta - u| = \tfrac{1}{2}|\alpha - \beta|$$

for both $u = n$ and $u = n + 1$. It would follow that

$$(\alpha - n)(n - \beta) = (n + 1 - \beta)(n + 1 - \alpha) = \tfrac{1}{2}(\alpha - \beta).$$

Also there would be equality in the earlier part of the argument. Now since $2\sqrt{cd} = c + d$ implies that $c = d$, we would have

$$(n - \beta)(n + 1 - \alpha) = (\alpha - n)(n + 1 - \beta) = \tfrac{1}{2}(\alpha - \beta).$$

A comparison of this and the preceding equations involving $\tfrac{1}{2}(\alpha - \beta)$ gives

$$\tfrac{1}{2}(\alpha - \beta) = (\alpha - n)(n - \beta) = (\alpha - n)(n + 1 - \beta).$$

Since $\alpha - n \neq 0$, we conclude that $n - \beta = n + 1 - \beta$, which is a contradiction.

It may be observed that this lemma cannot be improved without a change in form. For if $\alpha = \beta = \tfrac{1}{2}$, then (1) is best possible; and the constant $\tfrac{1}{2}$ cannot be improved in (2) in case $\alpha = \tfrac{1}{2}$ and β is large.

THEOREM 2.3. *Let* $a_1 x + b_1 y + c_1$ *and* $a_2 x + b_2 y + c_2$ *be linear forms with real coefficients such that* $\Delta = |a_1 b_2 - a_2 b_1| \neq 0$. *Furthermore, suppose that* a_1/b_1 *is irrational and that* $a_1 x + b_1 y + c_1 = 0$ *has no solutions in integers. Then for any given positive* ε, *there are infinitely many pairs of integers* x, y *such that*

$$(3) \qquad |a_1 x + b_1 y + c_1| \cdot |a_2 x + b_2 y + c_2| < \frac{\Delta}{4} \quad and$$

$$|a_1 x + b_1 y + c_1| < \varepsilon.$$

Proof: Since a_1/b_1 is irrational, then by Hurwitz' Theorem (with $\theta = -a_1/b_1$) there are infinitely many rational numbers h/k in lowest terms such that

$$(4) \qquad \left| -\frac{a_1}{b_1} - \frac{h}{k} \right| = \left| \frac{a_1}{b_1} + \frac{h}{k} \right| < \frac{1}{\sqrt{5}k^2}.$$

We may presume k positive so that

$$(5) \qquad |a_1 k + b_1 h| < \frac{|b_1|}{k\sqrt{5}}.$$

We restrict our attention to the infinitely many h/k which have k sufficiently large so as to satisfy

$$(6) \qquad \frac{|b_1|}{k\sqrt{5}} < \varepsilon \quad and \quad \left| \frac{b_1 b_2}{k^2} \right| < \Delta.$$

For any such h/k, choose r and s, that is, $r(h, k)$ and $s(h, k)$ such that

(7) $|ks - hr + \rho| \leqq \frac{1}{2}$,   where $\rho = \dfrac{k(c_1a_2 - c_2a_1) + h(c_1b_2 - c_2b_1)}{a_2b_1 - a_1b_2}$.

Lemma 2.1 assures us that this can be done.

Next we apply Lemma 2.2 with α and β defined by

$$\alpha = \frac{a_1r + b_1s + c_1}{a_1k + b_1h}, \qquad \beta = \frac{a_2r + b_2s + c_2}{a_2k + b_2h}.$$

(Discard any h, k for which $a_2k + b_2h = 0$, at most one pair.)   Write x for $r - uk$ and y for $s - uh$.   Then the condition $|\alpha - u| < 1$ in Lemma 2.2 becomes

(8) $|a_1x + b_1y + c_1| < |a_1k + b_1h|$

after multiplication by $|a_1k + b_1h|$.   Using (5) and (6) we obtain

(9) $|a_1x + b_1y + c_1| < \varepsilon$.

Now in case (1) holds in Lemma 2.2 we get, after multiplying by

$$|a_1k + b_1h| \quad \text{and} \quad |a_2k + b_2h|,$$

(10) $|\alpha_1x + b_1y + c_1| \cdot |a_2x + b_2y + c_2| \leqq \frac{1}{4}|a_1k + b_1h| \cdot |a_2k + b_2h|$.

Now by (4) we see that

$$\frac{h}{k} = -\frac{a_1}{b_1} + \frac{\delta}{k^2}, \qquad \text{where } |\delta| < \frac{1}{\sqrt{5}},$$

and hence

$|a_1k + b_1h| \cdot |a_2k + b_2h|$

$\leqq \dfrac{|b_1|}{k\sqrt{5}}|a_2k + b_2h| = \dfrac{|b_1|}{\sqrt{5}}\left|a_2 + b_2\dfrac{h}{k}\right|$

$= \dfrac{|b_1|}{\sqrt{5}}\left|a_2 + b_2\left(-\dfrac{a_1}{b_1} + \dfrac{\delta}{k^2}\right)\right| \leqq \dfrac{|b_1|}{\sqrt{5}}\left|a_2 + b_2\left(-\dfrac{a_1}{b_1}\right)\right| + \dfrac{|b_1|}{\sqrt{5}}\left|\dfrac{b_2\delta}{k^2}\right|$

$= \dfrac{\Delta}{\sqrt{5}} + \dfrac{|b_1b_2\delta|}{k^2\sqrt{5}} < \dfrac{\Delta}{\sqrt{5}} + \dfrac{\Delta}{\sqrt{5}} < \Delta.$

This with (10) implies (3), because of (9).

On the other hand, if (2) holds in Lemma 2.2 we get, after multiplying by $|a_1k + b_1h|$ and $|a_2k + b_2h|$,

$$(11) \quad |a_1x + b_1y + c_1| \cdot |a_2x + b_2y + c_2|$$

$$< \tfrac{1}{2}|(ks - hr + \rho)(a_2b_1 - a_1b_2)| = \frac{\Delta}{2}|ks - hr + \rho|$$

with ρ as in (7). Hence in view of (7) and (9) we have (3).

All that remains is to prove that there are infinitely many pairs x, y satisfying the inequalities (3). Suppose there were only a finite number of such pairs, say x_i, y_i with $1 \leq i \leq r$. Now by the hypotheses of the theorem $|a_1x_i + b_1y_i + c_1| \neq 0$ for $i = 1, 2, \cdots, r$. Thus we can choose h, k such that $|b_1|/(k\sqrt{5})$ is less than $|a_1x_i + b_1y_i + c_1|$ for all $i = 1, 2, \cdots, r$. Hence, in view of (5) and (8), the process would yield a pair x, y different from the r pairs x_i, y_i.

COROLLARY 2.4. *If θ is irrational and α is any real number such that $\theta x + y + \alpha = 0$ has no solution in integers x, y, then for any given positive ε there are infinitely many pairs of integers x, y such that $|x(\theta x + y + \alpha)| < \tfrac{1}{4}$ and $|\theta x + y + \alpha| < \varepsilon$.*

Proof: We apply Theorem 2.3 with a_1, b_1, c_1, a_2, b_2, c_2 replaced by θ, 1, α, 1, 0, 0 respectively, so that $\Delta = 1$. We conclude that there are infinitely many pairs of integers x, y satisfying $|x(\theta x + y + \alpha)| < \tfrac{1}{4}$ and $|\theta x + y + \alpha| < \varepsilon$.

THEOREM 2.5. *Let $a_1x + b_1y + c_1$ and $a_2x + b_2y + c_2$ be linear forms with real coefficients such that $\Delta = |a_1b_2 - a_2b_1| \neq 0$. Then there exists at least one pair of integers x, y such that*

$$(12) \quad |a_1x + b_1y + c_1| \cdot |a_2x + b_2y + c_2| \leq \frac{\Delta}{4}.$$

(There may be only a finite number of pairs of integers x, y satisfying the inequality. For example, consider the linear forms $x - \tfrac{1}{2}$ and $y - \tfrac{1}{2}$.)

Proof: In the case where a_1/b_1 is irrational and $a_1x + b_1y + c_1 = 0$ has no solution in integers, the result follows from Theorem 2.3. In case $a_1x + b_1y + c_1 = 0$ has a solution in integers, then these integers satisfy (12). There remain for consideration only those cases where neither a_1/b_1 nor a_2/b_2 is irrational. So a_1/b_1 is rational or $b_1 = 0$, and likewise a_2/b_2 is rational or $b_2 = 0$.

There is no loss in generality in assuming that a_1 and b_1 are relatively prime integers, for if $a_1x + b_1y + c_1$ is multiplied by any nonzero rational number r, $|\Delta|$ is replaced by $|\Delta r|$. Hence by Lemma 2.1 we can find integers x_0, y_0 such that

$$|a_1x_0 + b_1y_0 + c_1| \leqq \tfrac{1}{2}.$$

The same inequality would hold with x_0, y_0 replaced by $x_0 + b_1t$, $y_0 - a_1t$, where t is an arbitrary integer. Then we note that

$$a_2(x_0 + b_1t) + b_2(y_0 - a_1t) + c_2$$
$$= t(a_2b_1 - a_1b_2) + (a_2x_0 + b_2y_0 + c_2).$$

We choose t to be the integer nearest to

$$-\frac{a_2x_0 + b_2y_0 + c_2}{a_2b_1 - a_1b_2},$$

and hence we can conclude that

$$|a_1(x_0 + b_1t) + b_1(y_0 - a_1t) + c_1| \cdot |a_2(x_0 + b_1t) + b_2(y_0 - a_1t) + c_2|$$

$$\leqq \frac{1}{2} \cdot \frac{\Delta}{2} = \frac{\Delta}{4}.$$

This completes the proof.

2.2. *Further results*

Lemma 2.2 has been studied further by Eggan and Maier (1961) as follows. For real numbers α, β define $f(\alpha, \beta) = \min |\alpha - u| \cdot |\beta - u|$, where the minimum is to be taken over all integers u. Define

$$m(c) = \max_{|\alpha - \beta| = c} f(\alpha, \beta).$$

Eggan and Maier give an explicit determination of $m(c)$ as a function of c.

There are many proofs of Theorem 2.3, Corollary 2.4, and Theorem 2.5 in the literature; for example, see Pall (1943), Cassels (1957, Chapter

III), Hardy and Wright (1960, Chapter XXIV), and Niven (1961a). For a proof that the constant $\frac{1}{4}$ in Corollary 2.4 is best possible, see Cassels (1957, p. 48). In case the restriction $x > 0$ is imposed in Corollary 2.4 the problem becomes more difficult; for a review of this situation, see Cassels (1954). For conditions under which equality can be obtained in Theorem 2.5, see Hardy and Wright (1960, p. 403).

The following generalization of Theorem 2.5 is known as the Minkowski conjecture: If $L_j(x_1, \cdots, x_n)$ with $j = 1, \cdots, n$ are linear forms in $x_1, \cdots, x_n$ with determinant $\Delta \neq 0$, and if $c_1 \cdots c_n$ are any real numbers, then there are integers $x_1, \cdots, x_n$ such that

$$\prod_{j=1}^{n} |L_j(x_1, x_2, \cdots, x_n) + c_j| \leq \frac{|\Delta|}{2^n}.$$

This has been established for $n = 1, 2, 3, 4$ only; for details see Hardy and Wright (1960, p. 413).

CHAPTER 3

The Multiples of an Irrational Number

3.1. *A sequence of rational approximations to an irrational number*

For any irrational θ the theorems of Chapter 1 tell us something about the rational approximations h/k. The next theorem gives us some control over the sequence of denominators of such approximations.

THEOREM 3.1. *Given any irrational θ there is an increasing sequence of positive integers $k_1 < k_2 < k_3 < \cdots$ such that for $i \geqq 1$*

$$(1) \qquad \|k_i\theta\| = |k_i\theta - h_i| < \frac{1}{k_{i+1}}$$

Proof: For any distinct integers k and r note that $\|k\theta\| \neq \|r\theta\|$. Set $k_1 = 1$ and it follows that

$$\|\theta\| = \|k_1\theta\| = |k_1\theta - h_1| < \tfrac{1}{2}.$$

Define the integer m_2 by

$$\frac{1}{1 + m_2} < \|k_1\theta\| < \frac{1}{m_2},$$

so that $m_2 \geqq 2$. Let k_2 be the positive integer satisfying

$$\|k_2\theta\| = \min_{1 \leqq n \leqq m_2} \|n\theta\|,$$

so that by Theorem 1.1

$$\|k_2\theta\| < \frac{1}{1 + m_2} < \|k_1\theta\|, \qquad k_2 > k_1.$$

Also since $k_2 \leqq m_2$ we see that $\|k_1\theta\| < 1/m_2 \leqq 1/k_2$.

Now we use induction. With k_{i-1} obtained, we define m_i by

$$\frac{1}{1+m_i} < \|k_{i-1}\theta\| < \frac{1}{m_i}.$$

Then define k_i by

$$\|k_i\theta\| = \min_{1 \leq n \leq m_i} \|n\theta\|,$$

so that by Theorem 1.1

$$\|k_i\theta\| < \frac{1}{1+m_i} < \|k_{i-1}\theta\|, \qquad m_i \geq k_i > k_{i-1}.$$

Thus we see that

$$\|k_{i-1}\theta\| < \frac{1}{m_i} \leq \frac{1}{k_i},$$

and the theorem is proved.

3.2. *The uniform distribution of the fractional parts*

For any irrational θ, the fractional parts of the sequence θ, 2θ, 3θ, 4θ, $\cdots$ are not only dense in the unit interval but also uniformly distributed in the following sense. Let $\alpha_1, \alpha_2, \alpha_3, \cdots$ be any sequence of real numbers satisfying $0 \leq \alpha_i \leq 1$ for all i. Let I be any subinterval, open, closed or half-open, of the unit interval. Let I have length γ and let $N(I, n)$ denote the number of members of the finite sequence $\alpha_1, \alpha_2, \cdots, \alpha_n$ that lie in the interval I. Then the sequence is said to be uniformly distributed in the unit interval, or uniformly distributed modulo 1, if and only if

$$(2) \qquad \lim_{n \to \infty} \frac{N(I, n)}{u} = \gamma$$

for every subinterval I. In general, any sequence $\{\beta_i\}$ of real numbers is said to be uniformly distributed modulo 1 if the fractional parts $\beta_i - [\beta_i]$ are uniformly distributed in the unit interval.

THEOREM 3.2. *The sequence of fractional parts of the multiples of an irrational number* θ,

$$\theta - [\theta], \; 2\theta - [2\theta], \; 3\theta - [3\theta], \; \cdots$$

is uniformly distributed in the unit interval.

Proof: Let $\{k_i\}$ be a sequence as in Theorem 3.1. For any sufficiently large positive integer n, define i by

(3) $$k_i\sqrt{k_{i-1}} \leqq n < k_{i+1}\sqrt{k_i}.$$

Define $\delta = \delta(k_i)$ by

$$\theta = \frac{h_i}{k_i} + \frac{\delta}{k_i k_{i+1}},$$

so that $|\delta| < 1$ by Theorem 3.1. Dividing k_i into n we get a quotient q and remainder r; thus $n = qk_i + r$ with $0 \leqq r < k_i$.

Now any segment (open or closed) of length ρ of the real line covers at least $[\rho - 1]$ points with integer coordinates. Hence any segment of length γ covers at least $[\gamma k_i - 1]$ of the points whose coordinates are multiples of $1/k_i$. It follows that a subinterval I of length γ of the unit interval covers at least $[\gamma k_i - 1]$ of the points

(4) $$\frac{0}{k_i}, \frac{1}{k_i}, \frac{2}{k_i}, \cdots, \frac{k_i - 1}{k_i}.$$

By elementary congruence theory, the points (4) are the same in some order as

(5) $$(jh_i/k_i), \qquad j = 1, 2, \cdots, k_i,$$

where the parentheses denote fractional parts. The points

(6) $$(jh_i/k_i), \qquad j = 1, 2, \cdots, qk_i,$$

are simply the points (5) repeated q times. Hence we conclude that an interval I of length γ covers at least $q[\gamma k_i - 1]$ of the points (6).

Next, the distance between the points $j\theta$ and jh_i/k_i is

$$\left|\frac{j\delta}{k_i k_{i+1}}\right| < \frac{j}{k_i k_{i+1}} \leqq \frac{qk_i}{k_i k_{i+1}} \leqq \frac{n}{k_i k_{i+1}} < \frac{1}{\sqrt{k_i}},$$

by (3), where j is any integer in the set specified in (6). Hence the distance between the points $(j\theta)$ and (jh_i/k_i) is less than $1/\sqrt{k_i}$, provided we identify (if necessary) the endpoints 0 and 1 of the unit interval in computing distance. Thus the points

(7) $(j\theta)$, $j = 1, 2, \cdots, qk_i$

are the points (6), each translated by a distance less than $1/\sqrt{k_i}$. So our previous estimate of the number of points (6) in the interval I will carry over to (7), provided we delete a segment $1/\sqrt{k_i}$ at each end of the interval.   That is to say, an interval I of length γ covers at least

$$q\left[\left(\gamma - \frac{2}{\sqrt{k_i}}\right)k_i - 1\right]$$

points of (7). This estimate can still be used if we extend (7) to $j = 1, 2, \cdots, n$. So if $N(I, n)$ denotes the number of points among $(\theta), (2\theta), \cdots, (n\theta)$ that lie in I, we conclude that

$$\begin{aligned}
N(I, n) &\geq q\left[\left(\gamma - \frac{2}{\sqrt{k_i}}\right)k_i - 1\right] \\
&\geq q\left\{\left(\gamma - \frac{2}{\sqrt{k_i}}\right)k_i - 2\right\} \\
&= qk_i\left\{\gamma - \frac{2}{\sqrt{k_i}} - \frac{2}{k_i}\right\} \\
&\geq qk_i\left\{\gamma - \frac{4}{\sqrt{k_i}}\right\}.
\end{aligned}$$

Now since $n \geq qk_i > n - k_i$,

$$\frac{N(I, n)}{n} \geq \frac{qk_i\gamma}{n} - \frac{4}{\sqrt{k_i}} > \gamma - \frac{\gamma k_i}{n} - \frac{4}{\sqrt{k_i}},$$

$$\frac{N(I, n)}{n} - \gamma > - \frac{k_i}{n} - \frac{4}{\sqrt{k_i}}.$$

The terms on the right side of this inequality can be made arbitrarily small by taking n sufficiently large because of (3), and so given any positive ε we can conclude that

(8) $$\frac{N(I, n)}{n} - \gamma > - \frac{\varepsilon}{2}$$

for all sufficiently large n.   Let J be the complement of I in the unit interval; in general J consists of two intervals enclosing I between them. Applying (8) to these two intervals we obtain

$$(9) \qquad\qquad \frac{N(J, n)}{n} - (1 - \gamma) > -\varepsilon,$$

where $-\varepsilon/2$ in (8) has been replaced by $-\varepsilon$ because (9) is in general obtained by adding two inequalities of type (8).   Replacing $N(J, n)$ by $n - N(I, n)$ in (9) we get

$$\frac{N(I, n)}{n} - \gamma < \varepsilon,$$

and this together with (8) establishes the theorem.

3.3.   *The uniform distribution of the integral parts*

Consider the integral parts of the multiples of an irrational number θ, namely the sequence $[\theta]$, $[2\theta]$, $[3\theta]$, $\cdots$.   In the limiting sense are there just as many even numbers as odd, just as many in each of the three congruence classes modulo 3, and so on?   The answer is yes, as follows. Consider any sequence of integers b_1, b_2, b_3, $\cdots$.   For any modulus $m > 1$, define $N(n, j, m)$ as the number of integers b_k satisfying $k \leqq n$ and $b_k \equiv j \pmod{m}$.   Say that the sequence $\{b_i\}$ is *uniformly distributed modulo m* in case

$$\lim_{n \to \infty} \frac{N(n, j, m)}{n} = \frac{1}{m}$$

for each of the residue classes $j = 1, 2, 3, \cdots, m$.   Say that the sequence is *uniformly distributed* in case it is uniformly distributed modulo m for every positive integer $m \geqq 2$.

THEOREM 3.3.   *If θ is irrational then the integral parts of its multiples, $[\theta]$, $[2\theta]$, $[3\theta]$, $\cdots$, form a uniformly distributed sequence of integers.*

Proof: Applying Theorem 3.2 to the irrational θ/m, m being arbitrary but fixed, we see that the sequence

$$\left\{ \frac{k\theta}{m} - \left[\frac{k\theta}{m} \right] \right\}, \qquad k = 1, 2, 3, \cdots$$

is uniformly distributed in the unit interval. Multiplying the terms of this sequence by m we see that the sequence

$$\left\{ k\theta - m\left[\frac{k\theta}{m}\right]\right\}, \qquad k = 1, 2, 3, \cdots$$

is uniformly distributed (in the obvious sense) over the real line from 0 to m. Hence if we take the integral parts of this sequence, we get a sequence of integers which is uniformly distributed modulo m,

$$\left\{\left[k\theta - m\left[\frac{k\theta}{m}\right]\right]\right\} = \left\{[k\theta] - m\left[\frac{k\theta}{m}\right]\right\}, \qquad k = 1, 2, 3, \cdots.$$

However, modulo m we can ignore all multiples of m, and so we conclude that the sequence $\{[k\theta]\}$ is uniformly distributed modulo m. This is true for all $m \geq 2$, and so the theorem is proved.

Theorems 3.2 and 3.3 hold for any irrational. Of course, if θ satisfies $-1 < \theta < 1$ then the sequence of Theorem 3.3 does not consist of *distinct* integers, but the definition of a uniformly distributed sequence of integers contains no requirement that the integers be distinct.

3.4. *Kronecker's Theorem*

Besides the theorem of the preceding section, there are other results of interest concerning the integral parts of the multiples of an irrational. To obtain these results we need the work of the present section.

THEOREM 3.4. *Let α and β be irrational numbers such that 1, α, β are linearly independent over the field of rational numbers. Then the points whose coordinates are the fractional parts of the multiples of α and β,*

$$(10) \qquad (m\alpha - [m\alpha], m\beta - [m\beta]), \qquad m = 1, 2, 3, \cdots,$$

are dense in the unit square, i.e., given any real numbers λ_1, λ_2, μ_1, μ_2 satisfying $0 \leq \lambda_1 < \lambda_2 \leq 1$ and $0 \leq \mu_1 < \mu_2 \leq 1$ then there is a point (x, y) in the sequence (10) such that $\lambda_1 < x < \lambda_2$ and $\mu_1 < y < \mu_2$.

Proof: Let P_m be the point whose coordinates are shown in (10). Note that P_m is invariant under replacement of α and β by $\alpha + h$ and $\beta + k$, where h and k are arbitrary integers. Hence it suffices to prove the theorem for $0 < \alpha < 1$ and $0 < \beta < 1$. The points P_m, $m = 1, 2, 3,$

$\cdots$, are all distinct inside the unit square; no point is on the boundary of the square.

First we want to show that there are points P_k and P_n, with $n \neq k$, which are arbitrarily close. This can be done by the box principle. Consider the $m^2 + 1$ points P_k with $k = 1, 2, \cdots, m^2 + 1$. Dividing the unit square into m^2 subsquares of side $1/m$, we see that at least two of the points lie inside some subsquare. Thus two of the points are at a distance less than $\sqrt{2}/m$, which can be made arbitrarily small.

Define δ as the shortest distance from P_1 to the boundary of the unit square. Given any ε such that $\varepsilon < \delta$, let P_n and P_{n+r} be two points whose distance is less than ε. The directed path from P_n to P_{n+r} is a vector, denoted by $P_n P_{n+r}$. We note that $P_1 P_{1+r}$ is an equal vector in the sense that it is parallel to and of the same length as $P_n P_{n+r}$; this is so because we have restricted ε to satisfy $\varepsilon < \delta$, and hence there are no complications caused by the boundary of the unit square.

We now distinguish two cases according to whether there is or is

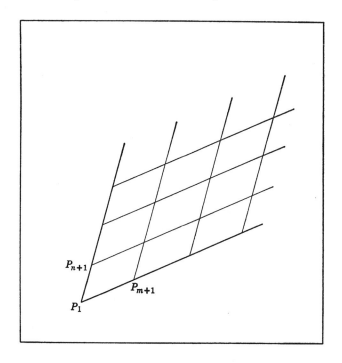

not a positive ε such that all vectors $P_n P_{n+r}$ of length $< \varepsilon$ are parallel. Taking first the case where there is no such positive ε, we have arbitrarily short nonparallel vectors, say $P_1 P_{m+1}$ and $P_1 P_{n+1}$, of length less than ε. Then we construct the lattice of parallelograms based on these two vectors, that is, all points $P_{1+xm+yn}$ with arbitrary nonnegative integers x and y. All vertices so obtained are points of the type (10), it being understood that all points reached by the lattice of parallelograms beyond the unit square are mapped back therein by suitable translations of integer length. This network of parallelograms covers the whole unit square; to see this, we note that if momentarily we do *not* map the network back into the unit square, but let it extend out into the plane, it covers some square (in fact, infinitely many squares) with integer coordinates which maps onto the entire unit square. Thus every point of the unit square is within ε of some point of type (10) for any given positive ε.

The other case, where for some ε all vectors of length $< \varepsilon$ are parallel, turns out to be impossible. There are infinitely many subscripts m such that $P_1 P_m$ has length $< \varepsilon$; we take two such subscripts n and r so that P_1, P_n, and P_r are collinear. Hence we have

$$\begin{vmatrix} \alpha & \beta & 1 \\ n\alpha - [n\alpha] & n\beta - [n\beta] & 1 \\ r\alpha - [r\alpha] & r\beta - [r\beta] & 1 \end{vmatrix} = 0,$$

$$\begin{vmatrix} \alpha & \beta & 1 \\ [n\alpha] & [n\beta] & n-1 \\ [r\alpha] & [r\beta] & r-1 \end{vmatrix} = 0,$$

$$a\alpha + b\beta + c = 0,$$

where a, b, c are integers. But 1, α, β are linearly independent, so $a = b = c = 0$. In particular, $b = 0$ implies

$$\frac{[r\alpha]}{r-1} = \frac{[n\alpha]}{n-1}.$$

Since this holds for infinitely many n and r, let us fix r and let n tend to infinity. The limit of $[n\alpha]/(n-1)$ is α, and so $\alpha = [r\alpha]/(r-1)$. But α is irrational, and so we have a contradiction.

A companion result to Theorem 3.4 will be useful in case 1, α, β are linearly dependent. This is obtained in Theorem 3.6. First, however, we prove a lemma.

LEMMA 3.5. *If* (λ_n, μ_n), $n = 1, 2, 3, \cdots$ *is a sequence of points which are dense on the line segment from* $(0, 0)$ *to* (h, k), *where* h *and* k *are relatively prime integers, then the points*

$$(\lambda_n + r, \mu_n + s),$$

$n = 1, 2, 3, \cdots;$ $r = 0, \pm 1, \pm 2, \pm 3, \cdots;$ $s = 0, \pm 1, \pm 2, \pm 3, \cdots,$

are dense on all the lines $kx - hy = t$, $t = 0, \pm 1, \pm 2, \pm 3, \cdots$.

Proof: We note that the points $(\lambda_n + h, \mu_n + k)$, $n = 1, 2, 3, \cdots$, are dense on the line segment from (h, k) to $(2h, 2k)$. Similarly the points $(\lambda_n + mh, \mu_n + mk)$, $n = 1, 2, 3, \cdots$, are dense on the line segment from (mh, mk) to $(mh + h, mk + k)$. For any given integer t_0 choose the integer pair x_0, y_0 so that $kx_0 - hy_0 = t_0$. Then the points

$$(\lambda_n + x_0 + mh, \mu_n + y_0 + mk), \qquad n = 1, 2, 3, \cdots$$

are dense on the line segment from

$$(x_0 + mh, y_0 + mk) \quad \text{to} \quad (x_0 + mh + h, y_0 + mk + k).$$

This is a segment of the line $kx - hy = t_0$, and all segments of this line arise by taking $m = 0, \pm 1, \pm 2, \pm 3, \cdots$.

THEOREM 3.6. *Let* α *and* β *be irrational but such that* 1, α, β *are linearly dependent over the field of rational numbers, say*

$$a\alpha + b\beta = c \qquad \text{with g.c.d. } (a, b, c) = 1.$$

Then the points (10) *whose coordinates are the fractonal parts of the multiples of* α *and* β *lie on, and only on, those portions of the lines* $ax + by = t$, *where* t *is any integer, lying within the unit square. Furthermore, the points* (10) *are dense on these segments.*

Proof: Since $a \neq 0$ and $b \neq 0$, we may presume that $b > 0$ because we can if necessary change the signs of a, b, and c. Define g as the greatest common divisor of a and b; thus $g = (a, b)$. Note that $(g, c) = 1$. Also there exist integers u and v such that

$$\frac{a}{g}u + \frac{b}{g}v = -c.$$

Then define

$$\alpha' = g\alpha + u, \qquad \beta' = g\beta + v,$$

so that α' and β' are irrational, and

(11) $$\frac{a}{g}\alpha' + \frac{b}{g}\beta' = 0.$$

Next define, for $n = 1, 2, 3, \cdots$,

$$w_n = \frac{gn\alpha'}{b} - \left[\frac{gn\alpha'}{b}\right].$$

By Theorem 3.2 the sequence $\{w_n\}$ is dense on the unit interval. Hence the sequence $\{bw_n/g\}$ is dense on the real line from 0 to b/g, and the points

$$\left(\frac{b}{g}w_n, \ -\frac{a}{g}w_n\right), \qquad n = 1, 2, 3, \cdots$$

are dense on the line segment from $(0, 0)$ to $(b/g, -a/g)$. Applying Lemma 3.5 we see that the points

(12) $$\left(\frac{b}{g}w_n + r, \ -\frac{a}{g}w_n + s\right),$$

$n = 1, 2, 3, \cdots$; $r = 0, \pm 1, \pm 2, \pm 3, \cdots$; $s = 0, \pm 1, \pm 2, \pm 3, \cdots$,

are dense on all the lines

(13) $$\frac{a}{g}x + \frac{b}{g}y = t, \qquad t = 0, \pm 1, \pm 2, \pm 3, \cdots.$$

Next, recalling that the notation $\theta \equiv \gamma \pmod 1$ means that $\theta - \gamma$ is an integer, we note that

$$\frac{b}{g}w_n \equiv \frac{b}{g} \cdot \frac{gn\alpha'}{b} \equiv n\alpha' \equiv ng\alpha \pmod 1,$$

$$-\frac{a}{g}w_n \equiv -\frac{a}{g} \cdot \frac{gn\alpha'}{b} \equiv n\beta' \equiv ng\beta \pmod 1,$$

by use of (11).   It follows that the points (12) can be written as

$$(n g \alpha + r, n g \beta + s),$$

$$n = 1, 2, 3, \cdots; \quad r = 0, \pm 1, \pm 2, \pm 3, \cdots; \quad s = 0, \pm 1, \pm 2, \pm 3, \cdots,$$

and so these are dense on the lines (13).

Let q be any given integer. Let t_0, m_0 satisfy $g t_0 + c m_0 = q$. Define T as the set of all triples (n, r, s) with

$$n = 1, 2, 3, \cdots, \quad r = 0, \pm 1, \pm 2, \pm 3, \cdots, \quad s = 0, \pm 1, \pm 2, \pm 3, \cdots.$$

Let T_1 denote the subset of T consisting of all triples (n, r, s) such that the points $(n g \alpha + r, n g \beta + s)$ lie on, and so are dense on,

$$\frac{a}{g} x + \frac{b}{g} y = t_0.$$

Translating each of these points by $m_0 \alpha$, $m_0 \beta$, we obtain the points

$$(n g \alpha + m_0 \alpha + r, \ n g \beta + m_0 \beta + s)$$

which satisfy

$$a(n g \alpha + m_0 \alpha + r) + b(n g \beta + m_0 \beta + s)$$

$$= a\{(n g + m_0)\alpha + r\} + b\{(n g + m_0)\beta + s\}$$

$$= g t_0 + a m_0 \alpha + b m_0 \beta = g t_0 + m_0 c = q.$$

Hence these points lie on $ax + by = q$, and are dense on this line.   But q is arbitrary, and hence the points

$$(m \alpha + r, \ m \beta + s),$$

$$m = 1, 2, 3, \cdots; \quad r = 0, \pm 1, \pm 2, \pm 3, \cdots; \quad s = 0, \pm 1, \pm 2, \pm 3, \cdots,$$

are dense on the lines $ax + by = t$, with $t = 0, \pm 1, \pm 2, \pm 3, \cdots$, because $a(m \alpha + r) + b(m \beta + s)$ is an integer for all m, r, s.   These lines include the segments in the unit square described in the theorem. Consequently the theorem follows at once if we take the fractional parts of the coordinates of $(m \alpha + r, \ m \beta + s)$.

3.5. *Results of Skolem and Bang*

For any positive real number α define N_α as the set of integers

$$\{[\alpha],\ [2\alpha],\ [3\alpha],\ \cdots,\ [n\alpha],\ \cdots\}.$$

As we shall see, these sets have many interesting properties.

THEOREM 3.7. *Let α and β be positive real numbers. Denote the set of all positive integers by Z and the null set by 0. Then $N_\alpha \cup N_\beta = Z$ and $N_\alpha \cap N_\beta = 0$ if and only if α and β are irrational and $\alpha^{-1} + \beta^{-1} = 1$.*

Remark. When the conditions $N_\alpha \cup N_\beta = Z$ and $N_\alpha \cap N_\beta = 0$ hold, we say that N_α and N_β are complementary sets of positive integers.

Proof: If $0 < \alpha \leqq 1$ then $N_\alpha = Z$ and $N_\alpha \cap N_\beta \neq 0$. Hence it suffices to restrict attention to $\alpha > 1$ and $\beta > 1$.

For any integer h let $\mu(\alpha, h)$ denote the number of elements of N_α that do not exceed h. Writing n for $\mu(\alpha, h)$ we see that

$$[n\alpha] \leqq h < [(n + 1)\alpha].$$

The first of these inequalities implies that $n\alpha - 1 < h$, and the second, being an inequality between integers, implies that

$$h + 1 \leqq [(n + 1)\alpha] \leqq (n + 1)\alpha.$$

Hence we conclude that

(14) $(h + 1)\alpha^{-1} - 1 \leqq n = \mu(\alpha, h) < (h + 1)\alpha^{-1},$

and so $\mu(\alpha, h)/h$ tends to α^{-1} as $h \to \infty$.

Now suppose that N_α and N_β are complementary sets. It follows that

(15) $\mu(\alpha, h) + \mu(\beta, h) = h$

for all h, and hence $\alpha^{-1} + \beta^{-1} = 1$. If the ratio α/β were rational, say $\alpha/\beta = a/b$, where a and b are positive integers, N_α and N_β would have the infinitely many common elements $[mb\alpha] = [ma\beta]$ for every integer $m \geqq 1$. Thus α/β is irrational, and this with $\alpha^{-1} + \beta^{-1} = 1$ implies that α and β are irrational.

Conversely, suppose that α and β are irrational and $\alpha^{-1} + \beta^{-1} = 1$. Then the equality sign cannot occur in the first inequality in (14), and so

$$(h + 1)\alpha^{-1} - 1 + (h + 1)\beta^{-1} - 1 < \mu(\alpha, h) + \mu(\beta, h)$$
$$< (h + 1)\alpha^{-1} + (h + 1)\beta^{-1},$$
$$h - 1 < \mu(\alpha, h) + \mu(\beta, h) < h + 1.$$

Consequently (15) holds for all h, and so N_α and N_β are complementary sets.

THEOREM 3.8. *Given positive real numbers α and β such that 1, α^{-1}, β^{-1} are linearly independent over the field of rational numbers, then N_α and N_β have infinitely many common elements.*

Proof: We may presume $\alpha > 1$ and $\beta > 1$ since $\alpha \leqq 1$ implies that $N_\alpha = Z$, the set of all positive integers. The hypotheses imply that α^{-1} and β^{-1} are irrational, and so we can apply Theorem 3.4 and conclude that there are infinitely many positive integers m such that

$$(16) \quad 0 < m\alpha^{-1} - [m\alpha^{-1}] < \alpha^{-1}, \qquad 0 < m\beta^{-1} - [m\beta^{-1}] < \beta^{-1}.$$

Writing h and k for $[m\alpha^{-1}]$ and $[m\beta^{-1}]$ we get

$$0 < m - h\alpha < 1, \qquad 0 < m - k\beta < 1,$$
$$[h\alpha] = [k\beta] = m - 1.$$

THEOREM 3.9. *Let α and β be positive irrational numbers such that $a\alpha^{-1} + b\beta^{-1} = c$ for some integers a, b, c with $ab < 0$ and $c \neq 0$. Then N_α and N_β have infinitely many common elements.*

Proof: We may presume that c is positive since the triplet a, b, c can be replaced by $-a$, $-b$, $-c$. As previously, we take $\alpha > 1$ and $\beta > 1$. Also, because of the symmetry of α and β we presume that $a > 0$ and $b < 0$. By Theorem 3.6 the points

$$(17) \quad (m\alpha^{-1} - [m\alpha^{-1}], \, m\beta^{-1} - [m\beta^{-1}]), \qquad m = 1, 2, 3, \cdots$$

are dense on the line $ax + by = 0$ in the unit square. Hence there are infinitely many positive integers m such that inequalities (16) hold as in the preceding theorem, and the rest of the argument follows as before.

THEOREM 3.10. *Let α and β be positive irrational numbers such that $a\alpha^{-1} + b\beta^{-1} = c$ for some positive integers a, b, c with g.c.d.$(a, b, c) = 1$ and $c > 1$. Then N_α and N_β have infinitely many common elements.*

Proof: Again we may presume that $\alpha > 1$ and $\beta > 1$. By Theorem 3.6, the points

(17) $(m\alpha^{-1} - [m\alpha^{-1}], \ m\beta^{-1} - [m\beta^{-1}])$, $m = 1, 2, 3, \cdots$

are dense on the line $ax + by = 1$ in the unit square. This line segment extends from $(0, b^{-1})$ to $(a^{-1}, 0)$. Since $c \geq 2$ we note that $a\alpha^{-1} > 1$ or $b\beta^{-1} > 1$, that is, $\alpha^{-1} > a^{-1}$ or $\beta^{-1} > b^{-1}$. Hence the line segment enters the rectangle defined by $0 < x < \alpha^{-1}, 0 < y < \beta^{-1}$, and so infinitely many of the points (17) satisfy the inequalities (16). The rest of the proof is the same as in Theorem 3.8.

THEOREM 3.11. *Let α and β be positive real numbers. The sets N_α and N_β are disjoint if and only if α and β are irrational and there exist positive integers a and b such that $a\alpha^{-1} + b\beta^{-1} = 1$. Furthermore, if N_α and N_β have one common element they have infinitely many.*

Proof: We can take $\alpha > 1$ and $\beta > 1$ since otherwise N_α and N_β have infinitely many common elements. If α and β are rational then there are infinitely many pairs of positive integers h, k such that $h\alpha = k\beta$, and so N_α and N_β have infinitely many common elements.

Next suppose that α is irrational but β is rational, say $\beta = q/r$, where q and r are positive integers. We prove that N_α and N_β have infinitely many common elements. By Theorem 3.1 (or Theorem 1.1), there are infinitely many positive integers $m > q$ such that $\|m\alpha\| < 1/m$, so that if k is the nearest integer to $m\alpha$,

(18) $0 < m\alpha - k < \dfrac{1}{m}$ or $0 < k - m\alpha < \dfrac{1}{m}.$

If the first of these inequalities holds, then

$$0 < qm\alpha - qk < \frac{q}{m} < 1,$$

$$[qm\alpha] = qk = kr\beta = [kr\beta],$$

so that qk belongs to both N_α and N_β. If the second inequality (18) holds, let t be the largest integer such that $t(k - m\alpha) < 1$. Then we get

$$(t + 1)(k - m\alpha) > 1, \qquad 0 < tm\alpha - (tk - 1) < k - m\alpha < \frac{1}{m},$$

$$0 < tqm\alpha - q(tk - 1) < \frac{q}{m} < 1,$$

$$[tqm\alpha] = q(tk - 1) = r\beta(tk - 1) = [r\beta(tk - 1)],$$

so that $q(tk - 1)$ belongs to both N_α and N_β.

Thus far we have completed the proof of Theorem 3.11 in case one or both of α, β are rational. Now let α and β be irrational. If $1, \alpha^{-1}, \beta^{-1}$ are linearly independent over the rationals, we use Theorem 3.8. If $1, \alpha^{-1}, \beta^{-1}$ are linearly dependent over the rationals, say $a\alpha^{-1} + b\beta^{-1} = c$ with $c \geqq 0$ and g.c.d.$(a, b, c) = 1$, Theorems 3.9 and 3.10 cover all cases with $c > 0$ except the case $c = 1, a > 0, b > 0$. Here we define $\gamma = a^{-1}\alpha, \rho = b^{-1}\beta$. Then $\gamma^{-1} + \rho^{-1} = 1$, and by Theorem 3.7 we observe that N_γ and N_ρ are disjoint. But N_α is a subset of N_γ, since $\alpha = a\gamma$, and likewise N_β is a subset of N_ρ. Hence N_α and N_β are disjoint.

Finally, there is the case $c = 0$, that is, $a\alpha^{-1} + b\beta^{-1} = 0$ for some integers a and b. Then for all integers m we have

$$[|b|m\alpha] = [|a|m\beta],$$

and the proof of Theorem 3.11 is complete.

THEOREM 3.12. *Let $\alpha > 1$ and $\beta > 1$ be irrational. Then $N_\alpha \cup N_\beta = Z$, the set of all positive integers, if and only if there are positive integers a and b such that*

$$a(1 - \alpha^{-1}) + b(1 - \beta^{-1}) = 1.$$

Proof: By Theorem 3.7 the complement of N_α is N_γ where γ is determined by $\alpha^{-1} + \gamma^{-1} = 1$. Similarly, the complement of N_β is N_ρ, where $\beta^{-1} + \rho^{-1} = 1$. Now $N_\alpha \cup N_\beta = Z$ if and only if N_γ and N_ρ are disjoint. Thus we apply Theorem 3.11 to N_γ and N_ρ, and this completes the proof.

THEOREM 3.13. *Let $\alpha > 1$ and $\beta > 1$ be irrational. Then $N_\alpha \supseteq N_\beta$ if and only if there are positive integers a and b such that $a(1 - \alpha^{-1}) + b\beta^{-1} = 1$.*

Proof: Using the notation of the previous theorem, we observe that $N_\alpha \supseteq N_\beta$ if and only if N_γ and N_β are disjoint. Thus we need merely apply Theorem 3.11 to N_γ and N_β.

THEOREM 3.14. *There are no positive real numbers α, β, γ such that N_α, N_β, N_γ are pairwise disjoint.*

Proof: If there were such real numbers α, β, γ, then by Theorem 3.11 there would be relations

$$a_1\alpha^{-1} + b_1\beta^{-1} = 1, \quad a_2\alpha^{-1} + b_2\gamma^{-1} = 1, \quad a_3\beta^{-1} + b_3\gamma^{-1} = 1,$$

with positive integral coefficients. The elimination of γ in the last two equations gives

$$a_2 b_3 \alpha^{-1} - a_3 b_2 \beta^{-1} = b_3 - b_2.$$

Now the coefficient $a_2 b_3$ is positive, whereas $-a_3 b_2$ is negative, so this equation gives a linear relation on α^{-1} and β^{-1} that is independent of the relation $a\alpha^{-1} + b\beta^{-1} = 1$. These two equations can be solved to give rational values for α and β, and this is not possible in view of Theorem 3.11.

3.6. *Sets defined by rational numbers*

The last few theorems have been concerned with sets N_α and N_β in case α and β are irrational. We now investigate what can be said about these sets in case one or both of α, β is rational. If α is rational, say $\alpha = r/s$ with $r > s > 0$ and $(r, s) = 1$, then N_α consists of the union of s arithmetic progressions A_j, namely

$$A_j = \{[j\alpha], \ r + [j\alpha], \ 2r + [j\alpha], \ 3r + [j\alpha], \ \cdots\},$$

for $j = 1, 2, 3, \cdots, s$. Stated in terms of residue classes, this means that N_α contains all positive integers from s of the residue classes modulo r, and no integers whatsoever from the other $r - s$ residue classes. For example, taking $j = s$ above, we see that N_α contains the entire set $r, 2r, 3r, 4r, \cdots$, all positive multiples of r. But N_α contains no integer x satisfying $x \equiv -1 \pmod{r}$ because if there were, there would have to be a value of j, as in A_j above, with $1 \leq j \leq s - 1$ such that

$$[j\alpha] \equiv -1 \pmod{r}, \qquad \left[\frac{jr}{s}\right] \equiv -1 \pmod{r}.$$

This is not possible because first if $s > 1$ then

$$1 \leq \left[\frac{r}{s}\right] < \left[\frac{2r}{s}\right] < \cdots < \left[\frac{(s-1)r}{s}\right] = r + \left[-\frac{r}{s}\right] < r - 1,$$

and second if $s = 1$ then every integer in N_α is a multiple of r.

Next suppose that β is irrational, $\beta > 1$. We now show that none of

(i) $N_\alpha \cup N_\beta = Z$,   (ii) $N_\alpha \cap N_\beta = 0$,

(iii) $N_\alpha \supseteq N_\beta$,   (iv) $N_\beta \supseteq N_\alpha$

can hold. For by Theorem 3.3, N_β contains infinitely many integers from every residue class modulo r. In view of the residue class of all multiples of r, we see that (ii) is impossible. Also the residue class of all positive integers $x \equiv -1 \pmod{r}$ shows that (iii) is impossible.

To see that (i) and (iv) are impossible we think of N_β as an increasing sequence of positive integers

$$[\beta], \ [2\beta], \ [3\beta], \ \cdots, \ [n\beta], \ \cdots,$$

and we use the concept of natural density. The natural density of an increasing sequence of integers $m_1, m_2, m_3, \cdots$ is the limit, if it exists, of $\mu(n)/n$ as $n \to \infty$, where $\mu(n)$ is the number of integers $\leq n$ in the sequence. It follows at once from this definition that N_β as a sequence has density β^{-1}. Furthermore, by Theorem 3.3 the subsequence of those integers in any given residue class modulo r has density $(\beta r)^{-1}$ because Theorem 3.3 says in effect that each residue class gets its proportionate share of members of N_β.

Consider the residue class of all positive integers $x \equiv -1 \pmod{r}$. The set N_α has no members in this class, and hence those members of $N_\alpha \cup N_\beta$ in this residue class, viewed as a sequence, have density $(\beta r)^{-1}$. This is less than r^{-1}, and hence (i) is impossible.

Consider the residue class of all positive multiples of r. As a sequence this has density r^{-1}. These integers are all members of N_α. But the subsequence of N_β made up of multiples of r has density $(\beta r)^{-1}$. Hence (iv) is impossible.

In case α and β are both rational, with $\alpha > 1$ and $\beta > 1$, a little more can be said about N_α and N_β. First we prove a partial analogue to Theorem 3.7.

THEOREM 3.15. *If ρ and σ are positive rational numbers such that $\rho^{-1} + \sigma^{-1} = 1$, then N_ρ and N_σ are almost complementary in the following sense. Let $\rho = k/m$ and $\sigma = k/(k - m)$ with $(k, m) = 1$, and let K denote the set of integers one less than those in N_k; thus*

$$N_k = \{jk\}, \qquad K = \{jk - 1\}, \quad j = 1, 2, 3, \cdots .$$

Then N_ρ and K are disjoint, N_σ and K are disjoint, and

$$(19) \qquad N_\rho \cap N_\sigma = N_k, \qquad N_\rho \cup N_\sigma \cup K = Z.$$

Proof: The analysis given in the proof of Theorem 3.7 is valid as far as (14). Thus with $\mu(\rho, h)$ defined as the number of integers in N_ρ that do not exceed h, and with n written in place of $\mu(\rho, h)$, we see that

$$n\rho < h + 1 \qquad \leqq (n + 1)\rho,$$

$$n < (h + 1)\rho^{-1} \leqq n + 1.$$

It follows that

$$\mu(\rho, h) = \begin{cases} (h + 1)\rho^{-1} - 1 & \text{if this is an integer,} \\ \\ [(h + 1)\rho^{-1}] & \text{otherwise.} \end{cases}$$

In case $(h + 1)\rho^{-1}$ is an integer, so is $(h + 1)\sigma^{-1}$ since $\rho^{-1} + \sigma^{-1} = 1$, and so we have

$$\mu(\rho, h) + \mu(\sigma, h) = (h + 1)\rho^{-1} - 1 + (h + 1)\sigma^{-1} - 1 = h - 1.$$

On the other hand, if $(h + 1)\rho^{-1}$ is not an integer, we have

$$\begin{aligned}
\mu(\rho, h) + \mu(\sigma, h) &= [(h + 1)\rho^{-1}] + [(h + 1)\sigma^{-1}] \\
&= [(h + 1)\rho^{-1}] + [h + 1 - (h + 1)\rho^{-1}] \\
&= h + 1 + [(h + 1)\rho^{-1}] + [-(h + 1)\rho^{-1}] \\
&= h.
\end{aligned}$$

Now by virtue of the notation set forth in the theorem, $(h + 1)\rho^{-1}$ is an integer if and only if $h = j(k - 1)$. The theorem follows by a simple analysis of sets.

Let us return to the general study of N_α and N_β, where α and β are rational, with $\alpha > 1$ and $\beta > 1$. These sets cannot be disjoint because if $\alpha = r/s$ and $\beta = m/n$, then $[ksm\alpha] = [krn\beta]$ for every positive integer k. It is also impossible that $N\alpha \cup N_\beta = Z$ because N_α contains no integer of the form $kr - 1$ and N_β contains no integer of the form $km - 1$, so that $N_\alpha \cup N_\beta$ contains no integer of the form $kmr - 1$.

However, it is possible that $N_\alpha \supseteq N_\beta$; in fact, the conditions for this are the same as in the irrational case in Theorem 3.13, as shown by the following result (Bang, 1957).

THEOREM 3.16. *Let $\alpha > 1$ and $\beta > 1$ be rational. Then $N_\alpha \supseteq N_\beta$ if and only if there are positive integers a and b such that*

$$(20) \qquad\qquad a(1 - \alpha^{-1}) + b\beta^{-1} = 1.$$

Proof: The sufficiency of the condition can be established from the irrational case. Let $\{\alpha_n\}$ be a decreasing sequence of irrationals with limit α. For n sufficiently large the sequence N_{α_n} coincides with N_α through a finite number of terms. Let the sequence $\{\beta_n\}$ be chosen to satisfy

$$a(1 - \alpha_n^{-1}) + b\beta_n^{-1} = 1,$$

so that it forms a decreasing sequence with limit β. Now if t is any integer in N_β we choose n sufficiently large so that N_{α_n} and N_α, and also N_{β_n} and N_β, coincide at least up to integers $\leq t$. Then we see by Theorem 3.13 that

$$t \in N_\beta, \qquad t \in N_{\beta_n}, \qquad t \in N_{\alpha_n}, \qquad t \in N_\alpha.$$

Conversely, assume that $N_\alpha \supseteq N_\beta$. First we establish that if $\alpha > 1$ is an integer, then β must also be an integer, and so β must be a multiple of α, thus giving (20) with $a = 1$ and $b = \beta\alpha^{-1}$. Suppose that $\beta = m/n$ with $(m, n) = 1$, $m > n > 1$. Choose a positive integer x to satisfy $mx \equiv 1 \pmod{n}$, say $mx = 1 + kn$. Then consider the members $[x\beta]$ and $[nx\beta]$ of N_β:

$$[x\beta] = k, \qquad [nx\beta] = mx, \qquad (k, mx) = 1,$$

and so it is not possible that all elements of N_β are multiples of the integer α.

Having completed the case where α is an integer, we presume in the rest of the proof that α is not an integer. Let α and β be written as fractions with a common numerator, say $\alpha = m/s$ and $\beta = m/n$, where g.c.d. $(m, n, s) = 1$. We show that g.c.d. $(m - s, n) = 1$. Writing d for this g.c.d., we note that m/d is a multiple of both m/n and $m/(m - s)$. It follows that since $N_\alpha \supseteq N_\beta$

$$N_{m/d} \subseteq N_{m/n} \subseteq N_{m/s}, \qquad N_{m/d} \subseteq N_{m/(m-s)},$$

(21) $$N_{m/d} \subseteq N_{m/s} \cap N_{m/(m-s)}.$$

Now m/s and $m/(m - s)$ can play the role of ρ and σ as in Theorem 3.15, with the property $\rho^{-1} + \sigma^{-1} = 1$. Hence by (19) the right side of (21) is N_k where $k = m/(m, s)$. Thus (21) becomes $N_{m/d} \subseteq N_k$, and so m/d is an integer, that is, d is a divisor of m. It follows that $d = 1$, since d is a divisor of m, $m - s$, and n.

The relation $N_\beta \subseteq N_\alpha$ states that to each positive integer j there corresponds an integer x_j such that $[j\beta] = [x_j\alpha]$. Thus

$$j\beta - y, \qquad x_j\alpha - y$$

have the same sign for all integers y (if the sign of zero is interpreted to be plus). Substituting for α and β, we conclude that

$$jm - yn, \qquad x_jm - ys$$

have the same sign for all j and all y. We use the absolute values of these numbers as a and b, with j and y chosen appropriately. First note that

$$a(1 - \alpha^{-1}) + b\beta^{-1} = \left| \left(jm - yn \right) \left(1 - \frac{s}{m} \right) + \left(x_j m - ys \right) \frac{n}{m} \right|$$

$$= |j(m - s) + (x_j - y)n|.$$

For any fixed positive integer j, note that $x_j - y$ can assume all integral values. Hence j and y can be determined so that the right side is the g.c.d. of n and $m - s$, namely 1.

This argument proves the theorem unless one or the other of these values of a and b is 0, and so we have two special cases to treat. First suppose that the value of b obtained by the preceding process is 0. Then some multiple of $1 - \alpha^{-1}$ equals 1, so that α has the form $r/(r - 1)$, and N_α consists of all positive integers except those congruent to -1 modulo r. Let $\beta = m/n$ with $(m, n) = 1$. Then $N_\alpha \supseteq N_\beta$ implies that for every positive integer j,

$$\left[\frac{jm}{n} \right] \not\equiv -1 \pmod{r}.$$

This expression says that

$$kr - 1 \leq \frac{jm}{n} < kr, \qquad krn - n \leq jm < krn, \qquad -n \leq jm - krn < 0$$

is impossible in positive integers j and k. Hence we see that

$$(m, rn) > n, \qquad (m, r) > n.$$

Then we choose

$$a = r - \frac{nr}{(m, r)}, \qquad b = \frac{m}{(m, r)},$$

and (20) is satisfied.

Finally, suppose that the value of a obtained by the preceding process is 0. Then β is an integer, say $\beta = q$, and write $\alpha = m/s$ with $(m, s) = 1$ and $m > s$. Then $N_\alpha \supseteq N_\beta$ implies that for any positive integer j there is a positive integer $x = x_j$ such that

$$\left[\frac{xm}{s} \right] = jq, \qquad jq \leq \frac{xm}{s} < jq + 1,$$

$$0 \leq xm - jqs < s, \qquad m \geq jqs - (x - 1)m > m - s.$$

It follows that

$$(m, qs) > m - s, \qquad (m, q) > m - s.$$

Then we choose

$$a = \frac{m}{(m, q)}, \qquad b = q - \frac{q(m - s)}{(m, q)},$$

and (20) is satisfied. This completes the proof.

3.7. Further results

Theorem 3.2 was first proved independently by Bohl, Sierpinski, and Weyl; see Koksma (1936, p. 92). Although the proof given in Section 3.2 avoids continued fractions by substituting Theorem 3.1, proofs by analytic techniques seem best suited to generalization. An exposition of such a proof by Fourier analysis is given in Niven (1956, Chapter 6). For the following n-dimensional generalization of Theorem 3.2, see Cassels (1957, Chapter 4): If $\theta_1 \cdots, \theta_n, 1$ are linearly independent over the rational numbers, then the points

$$(k\theta_1 - [k\theta_1], \ k\theta_2 - [k\theta_2], \ \cdots, \ k\theta_n - [k\theta_n])$$

for $k = 1, 2, 3, \cdots$ are uniformly distributed in the unit cube in n-space. This result gives at once a generalization of Theorem 3.3 as follows: With $\theta_1, \cdots, \theta_n$ as above, the n-tuples of integers

$$([k\theta_1], \ [k\theta_2], \ \cdots, \ [k\theta_n])$$

for $k = 1, 2, 3, \cdots$ are uniformly distributed in the sense that if $m_1 \geqq 2, m_2 \geqq 2, \cdots, m_n \geqq 2$ are any moduli, and $a_1, a_2, \cdots, a_n$ are any integers, then the fraction of the n-tuples satisfying

$$[k\theta_1] \equiv a_1 \,(\mathrm{mod}\ m_1), \ [k\theta_2] \equiv a_2 \,(\mathrm{mod}\ m_2), \ \cdots, \ [k\theta_n] \equiv a_n \,(\mathrm{mod}\ m_n),$$

has limiting value $(m_1 m_2 m_3 \cdots m_n)^{-1}$.

For other results on the uniform distribution of sequences on integers, see Niven (1961) and Uchiyama (1961).

In view of the foregoing remarks, it may be noted that Theorem 3.4 does not tell the full story; the points are not only dense but also uniformly distributed in the unit square. However, Theorem 3.4 as given, with a proof following Lettenmeyer (1923), is an adequate basis for the work of Skolem (1957) and Bang (1957) in Sections 3.5 and 3.6.

The proof of Theorem 3.6 was suggested to the writer by H. S. Zuckerman.

Theorem 3.7 is an often rediscovered result. It has been known for some time (see Beatty, 1926).

Let β satisfy $0 < \beta < 1$, and define R_β as the set of all pointwise integers m such that

$$[m\beta] < [(m + 1)\beta].$$

The study of such sets is included in this chapter because if $\alpha\beta = 1$, then $R_\beta = N_\alpha$.

CHAPTER 4

The Approximation of Complex Numbers

4.1. *Ford's theorem*

The results of this chapter are the analogues for complex numbers of the theorems on real numbers in Chapter 1. The central result, Theorem 4.3, is by L. R. Ford (1925); the proof given here follows that of O. Perron (1930, 1931). A complex integer, or Gaussian integer, is a number of the form $a + bi$, where a and b are rational integers. A rational complex number is one which can be expressed as the quotient of two complex integers; the field of rational complex numbers will be denoted by $R(i)$. An irrational complex number is one that is not rational; note that an irrational complex number $a + bi$ may have one rational component, a or b, but not both.

THEOREM 4.1. *Given any irrational complex number α there exist infinitely many rational complex numbers u/v such that*

$$\left| \alpha - \frac{u}{v} \right| < \frac{2}{|v|^2}.$$

Proof: Let v run through the $(n + 1)^2$ complex integers $a + bi$, where $0 \leqq a \leqq n$ and $0 \leqq b \leqq n$. To each v select the complex integer u so that $\alpha v - u = x + yi$ has the property $0 \leqq x < 1$, $0 \leqq y < 1$. The $x + yi$ are $(n + 1)^2$ distinct numbers since $\alpha \notin R(i)$. Divide the unit square (with corners at $0, 1, 1 + i, i$) into n^2 subsquares of side $1/n$. At least one subsquare contains two of the $\alpha v - u$, say $\alpha v_1 - u_1$ and $\alpha v_2 - u_2$. The distance between these is at most the length of the diagonal of the subsquare; thus

$$|(\alpha v_1 - u_1) - (\alpha v_2 - u_2)| < \frac{\sqrt{2}}{n}.$$

Here we have a strict inequality because only one of the $\alpha v - u$ (the one with $v = 0$) lies at a corner of a subsquare. By the method of choice we have $v_1 \neq v_2$ and $|v_1 - v_2| \leq n\sqrt{2}$. Hence we get

$$(1) \qquad \left| \alpha - \frac{u_1 - u_2}{v_1 - v_2} \right| < \frac{\sqrt{2}}{n|v_1 - v_2|} \leq \frac{2}{|v_1 - v_2|^2}.$$

When we write v for $v_1 - v_2$, u for $u_1 - u_2$,

$$\left| \alpha - \frac{u}{v} \right| < \frac{2}{|v|^2}.$$

If there were only a finite number of pairs of complex integers u, v satisfying this inequality we could choose n sufficiently large so that by (1)

$$\left| \alpha - \frac{u}{v} \right| < \frac{\sqrt{2}}{n|v|}$$

would yield one more fraction u/v because $n|v|$ can be made arbitrarily large. Consequently there are infinitely many pairs u, v, and the theorem is proved.

LEMMA 4.2. *Let a be any complex number. To each complex number z_1 there corresponds a homologous number z (that is, a number z such that $z - z_1$ is a complex integer) such that*

$$|z^2 - a|^2 \leq \tfrac{7}{16} + |a|^2,$$

with equality only in the two cases $a = \tfrac{3}{4}$, z_1 homologous to $i/2$, and $a = -\tfrac{3}{4}$, z_1 homologous to $\tfrac{1}{2}$.

Proof: Writing $a = \alpha + i\beta$, we begin with the case $\alpha \geq 0$. To each z_2 there correspond infinitely many homologous numbers $z = x + iy$ such that $-\tfrac{1}{2} < y \leq \tfrac{1}{2}$, and from these choose that one satisfying

$$(2) \qquad -\tfrac{1}{2} + (\tfrac{1}{4} - y^2)^{1/2} < x \leq \tfrac{1}{2} + (\tfrac{1}{4} - y^2)^{1/2}.$$

Then define

$$P = |z^2 - a|^2 - |a|^2, \qquad Q = |(z - 1)^2 - a|^2 - |a|^2,$$
$$R = |(z + 1)^2 - a|^2 - |a|^2.$$

We prove that at least one of P, Q, R is less than $\frac{7}{16}$, except in case $a = \frac{3}{4}$, $z = i/2$ when there is equality.

By simple algebra we see that

$$(3) \quad P = \{(x + iy)^2 - (\alpha + i\beta)\}\{(x - iy)^2 - (\alpha - i\beta)\}$$
$$- (\alpha + i\beta)(\alpha - i\beta),$$

and hence

$$(4) \qquad P = x^4 + 2x^2y^2 + y^4 - 2(x^2 - y^2)\alpha - 4xy\beta.$$

Also Q is the same with x replaced by $x - 1$, and R with x replaced by $x + 1$. It follows that

$$(5) \qquad (1 - x)P + xQ = u - 3u^2 + 2uv + v^2 + 2(v - u)\alpha,$$

where $u = x - x^2$ and $v = y^2$, and

$$(6) \quad (1 - x^2 - y^2)P + \tfrac{1}{2}(x^2 + y^2 + x)Q + \tfrac{1}{2}(x^2 + y^2 - x)R$$
$$= y^2 - 3x^2 + 3y^4 + 6y^2x^2 + 3x^4.$$

Separate the interval (2) into two parts:

$$(7) \qquad \tfrac{1}{2} - (\tfrac{1}{4} - y^2)^{1/2} < x \leq \tfrac{1}{2} + (\tfrac{1}{4} - y^2)^{1/2},$$

and

$$(8) \qquad -\tfrac{1}{2} + (\tfrac{1}{4} - y^2)^{1/2} < x \leq \tfrac{1}{2} - (\tfrac{1}{4} - y^2)^{1/2}.$$

In case (7) holds, we have $0 < x \leq 1$;

$$0 \leq (x - \tfrac{1}{2})^2 \leq \tfrac{1}{4} - y^2, \qquad x - x^2 \leq \tfrac{1}{4},$$
$$0 \leq y^2 \leq x - x^2 \leq \tfrac{1}{4}, \qquad 0 \leq v \leq u \leq \tfrac{1}{4}.$$

Then by (5) we see that

$$\min(P, Q) \leq (1 - x)P + xQ = u - 3u^2 + 2uv + v^2 + 2(v - u)\alpha$$
$$\leq u - 3u^2 + 2u^2 + u^2 = u \leq \tfrac{1}{4} < \tfrac{7}{16}.$$

In case (8) holds, we have $|x| \leq \tfrac{1}{2}$ and

$$(\tfrac{1}{2} \pm x)^2 \geq \tfrac{1}{4} - y^2, \qquad x^2 + y^2 \pm x \geq 0$$

for both signs. Also we have $1 - x^2 - y^2 \geq 0$ since $|x| \leq \tfrac{1}{2}$ and

$|y| \leqq \frac{1}{2}$. Hence in (6) the multipliers of P, Q, R are nonnegative with sum 1, so that

$\min (P, Q, R)$

$$\leqq (1 - x^2 - y^2)P + \tfrac{1}{2}(x^2 + y^2 + x)Q + \tfrac{1}{2}(x^2 + y^2 - x)R$$
$$= y^2 - 3x^2 + 3y^4 + 6y^2x^2 + 3x^4$$
$$\leqq \tfrac{1}{4} - 3x^2 + \tfrac{3}{16} + \tfrac{3}{2}x^2 + 3x^4$$
$$= \tfrac{7}{16} - \tfrac{3}{2}x^2 + 3x^4 = \tfrac{7}{16} - \tfrac{3}{2}x^2(1 - 2x^2).$$

Since $1 - 2x^2$ is positive, we conclude that $\min (P, Q, R) \leqq \tfrac{7}{16}$ with equality possible only in case $x = 0$, $y = \tfrac{1}{2}$, $z = x + iy = i/2$. In this case (6) becomes

$$\tfrac{3}{4}P + \tfrac{1}{8}Q + \tfrac{1}{8}R = \tfrac{7}{16}.$$

It follows that $\min (P, Q, R) = \tfrac{7}{16}$ if and only if $P = Q = R$ with $z = i/2$; thus

$$\left| \left(\frac{i}{2}\right)^2 - a \right| = \left| \left(\frac{i}{2} - 1\right)^2 - a \right| = \left| \left(\frac{i}{2} + 1\right)^2 - a \right|.$$

Geometrically, this means that the point a in the complex plane is equally distant from $-\tfrac{1}{4}$, $\tfrac{3}{4} - i$, and $\tfrac{3}{4} + i$. Thus $a = \tfrac{3}{4}$. This completes the proof of Lemma 4.2 in case $\alpha \geqq 0$.

In case $\alpha < 0$ we apply what has already been proved to $-a$ and iz_1. Thus we conclude that there is a number z_0 homologous to iz_1 such that

$$|z_0^2 + a|^2 \leqq \tfrac{7}{16} + |-a|^2 = \tfrac{7}{16} + |a|^2.$$

When we write z in place of $-iz_0$, we see that z is homologous to z_1 and also

$$|z^2 - a|^2 = |(-iz_0)^2 - a|^2 = |z_0^2 + a|^2 \leqq \tfrac{7}{16} + |a|^2.$$

Finally, there is equality here only if $-a = \tfrac{3}{4}$ and iz_1 is homologous to $i/2$; thus $a = -\tfrac{3}{4}$ and z_1 is homologous to $\tfrac{1}{2}$.

THEOREM 4.3. *Given any irrational complex number* α *there exist infinitely many rational complex numbers* u/v *such that*

$$\left| \alpha - \frac{u}{v} \right| < \frac{1}{\sqrt{3}|v|^2}.$$

Furthermore the constant $\sqrt{3}$ is best possible in the sense that the result becomes false for any larger constant.

Proof: Consider any rational complex number u/v in lowest terms, with $|v| > 1$. Define δ by the equation

$$\alpha - \frac{u}{v} = \frac{\delta}{v^2}.$$

Note that $\delta \neq 0$; in fact, δ is irrational. To each u/v there correspond infinitely many pairs of complex integers u_1, v_1 satisfying the equation $uv_1 - vu_1 = 1$. [This follows from the unique factorization property of complex integers (Niven and Zuckerman, 1960, p. 193); the theory of linear Diophantine equations in complex integers is directly analogous to the corresponding theory in rational integers.] Any solution $u_1 = \theta$, $v_1 = \rho$ of this equation generates them all: $u_1 = \theta + tu$, $v_1 = \rho + tv$, where t is an arbitrary complex integer.

Now we apply Lemma 4.2 with $z_1 = \rho/v + 1/(2\delta)$ and $a = 1/(4\delta^2)$. Hence there is a value of t yielding $z = v_1/v + 1/(2\delta)$ such that

$$(9) \qquad \left|\left(\frac{v_1}{v} + \frac{1}{2\delta}\right)^2 - \frac{1}{4\delta^2}\right| < \sqrt{\frac{7}{16} + \frac{1}{16|\delta|^4}}.$$

This is a strict inequality because $\rho/v + 1/(2\delta)$ cannot be homologous to $i/2$ or $1/2$ because δ is irrational. Define δ_1 by the equation

$$\alpha - \frac{u_1}{v_1} = \frac{\delta_1}{v_1^2},$$

and we see that

$$\alpha = \frac{u_1}{v_1} + \frac{\delta_1}{v_1^2} = \frac{u}{v} + \frac{\delta}{v^2},$$

$$\delta_1 = v_1^2\left(\frac{u}{v} - \frac{u_1}{v_1} + \frac{\delta}{v^2}\right) = \delta\left\{\left(\frac{v_1}{v} + \frac{1}{2\delta}\right)^2 - \frac{1}{4\delta^2}\right\}.$$

Then using (9) we conclude that

$$(10) \qquad |\delta_1|^2 = |\delta|^2 \cdot \left|\left(\frac{v_1}{v} + \frac{1}{2\delta}\right)^2 - \frac{1}{4\delta^2}\right| < \frac{7}{16}|\delta|^2 + \frac{1}{16|\delta|^2}.$$

Now we specialize the u/v in the foregoing analysis to be the infinitely many rational complex numbers from Theorem 4.1, omitting the finite number of these having $|v| = 1$. Note that for each such u/v the inequality $|\delta| < 2$ is satisfied. Furthermore, each such u/v yields a rational complex number u_1/v_1, and we prove first that the process gives infinitely many u_1/v_1. We do this by showing that a specific u_1/v_1 cannot arise from infinitely many u/v. For the preceding analysis gives

$$\frac{\delta_1}{v_1^2} = \frac{u}{v} - \frac{u_1}{v_1} + \frac{\delta}{v_2} = \frac{1}{vv_1} + \frac{\delta}{v^2},$$

$$\left| \frac{\delta_1}{v_1^2} \right| = \left| \frac{1}{vv_1} + \frac{\delta}{v^2} \right| \leqq \left| \frac{1}{vv_1} \right| + \left| \frac{\delta}{v^2} \right|.$$

It is impossible that this relation is satisfied for infinitely many u/v because $|\delta_1/v_1^2|$ is a fixed positive number, but $1/|vv_1|$ and $|\delta/v^2|$ are arbitrarily small as $|v| \to \infty$.

Now by (10) together with $|\delta| < 2$ we observe that for any δ satisfying $2 \leqq |\delta|^2 < 4$,

$$|\delta_1^2| < \frac{7}{16}|\delta|^2 + \frac{1}{16|\delta|^2} \leqq \max_{2 \leqq x \leqq 4} \left(\frac{7}{16}x + \frac{1}{16x} \right)$$

$$= \frac{7}{16}(4) + \frac{1}{16.4} < 2.$$

Thus from any u/v with $2 \leqq |\delta|^2 < 4$ we get u_1/v_1 with $|\delta_1|^2 < 2$. Any u/v with $|\delta|^2 < 2$ is not treated by the preceding process. In either case we get $|\delta_1|^2 < 2$ or $|\delta|^2 < 2$.

Changing notation we now have infinitely many u/v with $|\delta|^2 < 2$. We iterate the process. Any such δ with $1 \leqq |\delta|^2 < 2$ gives, by (10), a δ_1 satisfying

$$|\delta_1|^2 < \frac{7}{16}|\delta|^2 + \frac{1}{16|\delta|^2} \leqq \max_{1 \leqq x \leqq 2} \left(\frac{7}{16}x + \frac{1}{16x} \right)$$

$$= \frac{7}{16}(2) + \frac{1}{16.2} < 1.$$

Hence we have infinitely many u/v with $|\delta|^2 < 1$. Repeating the process, we apply (10) to any δ with $\frac{1}{2} \leqq |\delta|^2 < 1$ to get

$$|\delta_1|^2 < \max_{\frac{1}{2} \leqq x \leqq 1} \left(\frac{7}{16} x + \frac{1}{16x}\right) = \tfrac{7}{16} + \tfrac{1}{16} = \tfrac{1}{2}.$$

Next, to any δ with $\frac{3}{7} \leqq |\delta|^2 < \frac{1}{2}$ we get

$$|\delta_1|^2 < \max_{\frac{3}{7} \leqq x \leqq \frac{1}{2}} \left(\frac{7}{16} x + \frac{1}{16x}\right) = \tfrac{7}{16}(\tfrac{1}{2}) + \tfrac{2}{16} < \tfrac{3}{7}.$$

Finally, to any δ with $\frac{1}{3} \leqq |\delta|^2 < \frac{3}{7}$ we get

$$|\delta_1|^2 < \max_{\frac{1}{3} \leqq x \leqq \frac{3}{7}} \left(\frac{7}{16} x + \frac{1}{16x}\right)$$

$$= \max\left(\frac{7}{16}\cdot\frac{1}{3} + \frac{3}{16}, \frac{7}{16}\cdot\frac{3}{7} + \frac{7}{16.3}\right) = \max\left(\tfrac{1}{3}, \tfrac{1}{3}\right) = \tfrac{1}{3}.$$

Note that no further iteration of the process is possible, and we have now proved the first part of Theorem 4.3.

To show that $\sqrt{3}$ is best possible, consider the number $\alpha = (1 + i\sqrt{3})/2$, and so

$$\frac{1 + i\sqrt{3}}{2} - \frac{u}{v} = \frac{\delta}{v^2}, \qquad \frac{i\sqrt{3}}{2} v - \frac{\delta}{v} = u - \frac{v}{2}.$$

Squaring the last equation and rearranging, we get

$$\frac{\delta^2}{v^2} - i\delta\sqrt{3} = u^2 - uv + v^2.$$

Now $u^2 - uv + v^2 \neq 0$ because $u^2 - uv + v^2 = 0$ implies that $u/v = (1 \pm i\sqrt{3})/2$, whereas u/v is complex rational. Hence $|u^2 - uv + v^2| \geqq 1$, and

$$\left|\frac{\delta^2}{v^2} - i\delta\sqrt{3}\right| \geqq 1,$$

$$\left|\frac{\delta^2}{v^2}\right| + |i\delta\sqrt{3}| \geqq 1.$$

Now suppose that there are infinitely many such approximations u/v with $|\delta| < 1/c$ for some fixed constant $c > \sqrt{3}$. Then we conclude that

$$\frac{1}{c^2} \cdot \frac{1}{|v|^2} + \frac{\sqrt{3}}{c} > 1,$$

$$\frac{1}{c^2|v|^2} > 1 - \frac{\sqrt{3}}{c} = \frac{c - \sqrt{3}}{c},$$

$$|v|^2 < \frac{1}{c(c - \sqrt{3})}.$$

But this inequality is satisfied by only a finite number of complex integers v.

Theorem 4.3 implies the following result.

COROLLARY 4.4.   *Given any complex numbers* $\alpha_1, \alpha_2, \beta_1, \beta_2$ *with* $\Delta \neq 0$, *where* $\Delta = |\alpha_1\beta_2 - \alpha_2\beta_1|$, *and given any positive* ε, *there are infinitely many pairs of complex integers* h, k *such that*

$$|\alpha_1 k + \beta_1 h| \cdot |\alpha_2 k + \beta_2 h| < \frac{\Delta}{\sqrt{3}} + \varepsilon.$$

The proof of this is directly analogous to the proof of Corollary 1.6.

4.2.   *Further results*

Hofreiter (1952) has proved that given any irrational complex number α there exists at least one rational complex number u/v such that

$$\left| \alpha - \frac{u}{v} \right| \leq \frac{1}{\sqrt{2 + \sqrt{3}}\,|v|^2}.$$

The constant here is best possible.

Corollary 4.4 can be improved by the removal of the ε, but without a strict inequality; see Perron (1932) and Chalk (1955).

CHAPTER 5

The Product of Complex Linear Forms

The purpose of this chapter is to establish the analogues in the complex case of the results of Chapter 2. The proofs can be made by analogy, provided we first establish a result paralleling Lemma 2.2, which is given as Theorem 5.3.

5.1. *The covering of lattice points*

A lattice point in the plane is any point (r, s) with integer coordinates.

THEOREM 5.1. *A rectangle of dimensions a by b (say with $a \leqq b$) has at least one lattice point in its interior or on its boundary no matter where it lies in the plane if and only if $a \geqq 1$ and $b \geqq \sqrt{2}$.*

Proof: If $a < 1$, then the rectangle can be placed between an axis of the coordinate system and a parallel line at a distance of one unit ($x = 0$ and $x = 1$ in standard notation) in such a way as to cover no lattice point. This can be done by placing the rectangle so that its sides are parallel to the axes.

If $a \geqq 1$ and $b < \sqrt{2}$, thus $1 \leqq a \leqq b < \sqrt{2}$, then the rectangle can be placed with sides inclined at 45° to the axes so that it lies completely inside the rectangle with vertices $(\frac{3}{2}, \frac{1}{2})$, $(\frac{1}{2}, \frac{3}{2})$, $(-\frac{1}{2}, \frac{1}{2})$, $(\frac{1}{2}, -\frac{1}{2})$. The rectangle so placed will cover no lattice point. Hence the conditions are necessary.

Conversely, consider a rectangle of dimensions a by b, with $a \geqq 1$ and $b \geqq \sqrt{2}$. To prove that such a rectangle covers a lattice point no matter how it is placed in the plane, it suffices to establish the result with $a = 1$ and $b = \sqrt{2}$. First, if the rectangle is placed with sides parallel to the coordinate axes, the conclusion is obtained at once.

So consider the case where the sides of the rectangle are not parallel to the axes. By translating the position of the rectangle through integer lengths in the directions of the axes, we may presume that the rectangle is in a position so that it is not intersected by either axis or by the lines $y = \pm x$. By an interchange of the axes if necessary, and a reversal of the positive and negative directions if necessary, we may presume that the sides of length $\sqrt{2}$ of the rectangle are lines with positive slope between 0 and 1. Note that these changes do not alter the question of whether the rectangle covers a lattice point.

Thus we may presume that the sides of the rectangle of length $\sqrt{2}$, being a distance 1 apart, are segments of some lines

(1) $f(x, y) = x \sin \alpha - y \cos \alpha - c = 0,$

(2) $g(x, y) = f(x, y) - 1 = 0,$

with

$$0 < \tan \alpha \leq 1, \qquad 0 < \alpha \leq \frac{\pi}{4}.$$

There are infinitely many lattice points lying on or between these lines; this is so because any line parallel to the x-axis has a segment of length $1/\sin \alpha$ between lines (1) and (2), and $1/\sin \alpha > 1$. Let (r, s) be any such lattice point, so that

$$f(r, s) \geq 0, \qquad g(r, s) \leq 0.$$

We prove that at least one of the points $(r + 1, s)$ and $(r + 1, s + 1)$ also lies between, or on one of, the lines (1) and (2). Suppose that $(r + 1, s)$ does not have this property; then we prove that $(r + 1, s + 1)$ does. We are supposing that

$$f(r + 1, s) > 0 \quad \text{and} \quad g(r + 1, s) > 0.$$

Then we see that

$$f(r + 1, s + 1) = f(r + 1, s) - \cos \alpha$$
$$= g(r + 1, s) + (1 - \cos \alpha) > 0,$$

and

$$g(r + 1, s + 1) = g(r, s) + \sin \alpha - \cos \alpha \leq 0$$

because $\sin \alpha \leq \cos \alpha$ for the values of α under consideration.

Thus from a lattice point (r, s) lying between or on one of the lines (1) and (2) we get another such lattice point, $(r + 1, s)$ or $(r + 1, s + 1)$ or both.   Repeating the process we obtain an infinite sequence of such lattice points.   Furthermore, it is an infinite sequence in both directions, for if (r, s) is such a lattice point so also is at least one of the points $(r - 1, s)$ and $(r - 1, s - 1)$.   Consider the set of lines, perpendicular to (1) and (2), through each of the points in this infinite sequence. If we prove that adjacent lines in this set are at a distance $\leq \sqrt{2}$, this will prove the theorem.   It suffices to consider two cases, first where adjacent lines pass through (r, s) and $(r + 1, s)$, and second where adjacent lines pass through (r, s) and $(r + 1, s + 1)$.   In the first case the lines are

$$x \cos \alpha + y \sin \alpha - r \cos \alpha - s \sin \alpha = 0$$

and

$$x \cos \alpha + y \sin \alpha - r \cos \alpha - s \sin \alpha - \cos \alpha = 0.$$

The distance between these lines is $\cos \alpha$, and we note that $\cos \alpha < \sqrt{2}$. In the second case the lines are

$$x \cos \alpha + y \sin \alpha - r \cos \alpha - s \sin \alpha = 0$$

and

$$x \cos \alpha + y \sin \alpha - r \cos \alpha - s \sin \alpha - \cos \alpha - \sin \alpha = 0.$$

The distance between these lines is $\cos \alpha + \sin \alpha$, and we note that the maximum value of $\cos \alpha + \sin \alpha$, for all α, is $\sqrt{2}$.

5.2. An inequality in the complex plane

We now prove an analogue of Lemma 2.1.

LEMMA 5.2.   *Given any complex integers* h, k *with greatest common divisor* 1, *and given any complex number* ρ, *there exist complex integers* r *and* s *such that* $|ks - hr + \rho| \leq 1/\sqrt{2}$.

Proof: The proof is analogous to that of Lemma 2.1.   Since h and k are relatively prime, the set $\{ks - hr\}$ runs through all complex integers as s and r run through all complex integers. Hence s and r can be chosen so that $ks - hr$ is the complex integer nearest to $-\rho$, and the result follows.

THEOREM 5.3. *If α and β are any complex numbers, there is a complex integer u such that $|\alpha - u| < 2$ and*

(3) $$|\alpha - u| \cdot |\beta - u| \leqq \tfrac{3}{4} \qquad \text{if } |\alpha - \beta| < \sqrt{2},$$

(4) $$|\alpha - u| \cdot |\beta - u| < \frac{|\alpha - \beta|}{\sqrt{2}} \qquad \text{if } |\alpha - \beta| \geqq \sqrt{2}.$$

Remark. This is not an exact analogue of Lemma 2.2 because in inequality (3) the constant $\tfrac{3}{4}$ is perhaps not best possible. However, (4) is best possible because if $\alpha = (1 + i)/2$ and $|\beta|$ is large, the inequality cannot be improved.

Proof: If $\alpha = \beta$, we can take u as the complex integer nearest to α, so that

$$|\alpha - u| = |\beta - u| \leqq \frac{1}{\sqrt{2}}, \ |\alpha - u| \cdot |\beta - u| \leqq \tfrac{1}{2} < \tfrac{3}{4},$$

and the case is finished. Henceforth we presume that $\alpha \neq \beta$.

We look upon α and β not only as complex numbers but also as two distinct points in the complex plane. Similarly, u is to be selected as a complex integer, and from a geometric viewpoint we think of u as a point chosen from among the lattice points, that is, points with integer coordinates. We now choose a new coordinate system. This will not affect the location of the points α, β, u, but their coordinates will in general be changed. Note for a fixed u the invariance of $\alpha - u$, $\beta - u$, and $\alpha - \beta$ under a change of coordinate system in which the unit of length is the same.

Impose a new coordinate system so that the real axis passes through the points α and β, with the origin midway between α and β, and with α on the positive end of the real axis. Thus if $2k$ denotes the distance between the points α and β, the coordinate system is chosen so that α has new coordinates $(k, 0)$ and β has new coordinates $(-k, 0)$. Henceforth in the proof the complex numbers belonging to the points α, β, u

will be in the new coordinate system. Thus $\alpha = k$, $\beta = -k$, and we write $u = x + yi$, where x and y are now not necessarily integers. However, we shall still refer to the points u as lattice points, even though their coordinates are perhaps not integers. In view of these conventions we can write

$$(5) \qquad |(\alpha - u)(\beta - u)|^2 = \{(x - k)^2 + y^2\}\{(x + k)^2 + y^2\}$$
$$= (x^2 + k^2 + y^2)^2 - 4k^2x^2.$$

CASE 1. $k^2 \geqq \frac{1}{2}$. Choose $u = (x, y)$ as the lattice point nearest to $((k^2 - \frac{1}{2})^{1/2}, 0)$. Since there is a lattice point within $1/\sqrt{2}$ of any point, we see that

$$\{x - (k^2 - \tfrac{1}{2})^{1/2}\}^2 + y^2 \leqq \tfrac{1}{2},$$
$$(6) \qquad x^2 + k^2 + y^2 \leqq 1 + 2x(k^2 - \tfrac{1}{2})^{1/2}.$$

Now since $2rs \leqq r^2 + s^2$ we see that

$$2x(k^2 - \tfrac{1}{2})^{1/2} \leqq x^2 + k^2 - \tfrac{1}{2},$$
$$\{1 + 2x(k^2 - \tfrac{1}{2})^{1/2}\}^2 = 1 + 4x^2(k^2 - \tfrac{1}{2}) + 4x(k^2 - \tfrac{1}{2})^{1/2}$$
$$\leqq 1 + 4x^2(k^2 - \tfrac{1}{2}) + 2x^2 + 2k^2 - 1$$
$$= 2k^2 + 4x^2k^2.$$

This together with (6) and (5) implies that

$$|(\alpha - u)(\beta - u)|^2 = (x^2 + y^2 + k^2)^2 - 4k^2x^2$$
$$\leqq \{1 + 2x(k^2 - \tfrac{1}{2})^{1/2}\}^2 - 4k^2x^2$$
$$\leqq 2k^2 + 4k^2x^2 - 4k^2x^2$$
$$= 2k^2 = \tfrac{1}{2}|\alpha - \beta|^2.$$

Next we show that

$$|(\alpha - \mu)^2(\beta - \mu)^2| = \tfrac{1}{2}|\alpha - \beta|^2$$

is not possible, and so we will have (4). This equality would occur only if equality occurred in (6) and in the result immediately after (6); thus

$$2x(k^2 - \tfrac{1}{2})^{1/2} = x^2 + k^2 - \tfrac{1}{2}; \qquad \text{so} \quad x = (k^2 - \tfrac{1}{2})^{1/2}.$$

But there is equality in (6) only if the point

$$((k^2 - \tfrac{1}{2})^{1/2},\ 0)$$

lies at the center of a lattice square, in which case there would be four choices for (x, y), and so at least two choices for x. Thus we could choose x to be other than $(k^2 - \tfrac{1}{2})^{1/2}$.

Furthermore, we see that

$$|\alpha - u| \leqq |u - (k^2 - \tfrac{1}{2})^{1/2}| + |(k^2 - \tfrac{1}{2})^{1/2} - \alpha|$$

$$\leqq \frac{1}{\sqrt{2}} + k - (k^2 - \tfrac{1}{2})^{1/2}$$

$$\leqq \frac{1}{\sqrt{2}} + k - \left(k - \frac{1}{\sqrt{2}}\right) = \sqrt{2}.$$

CASE 2. $k^2 < \tfrac{1}{2}$. Apply Theorem 5.1 to obtain a lattice point $u = (x, y)$ inside or on the boundary of the rectangle defined by $|x| \leqq 1/\sqrt{2}$, $|y| \leqq 1/2$. Thus we have

$$|\alpha - u| \leqq |\alpha| + |u| \leqq k + \sqrt{\tfrac{3}{4}} \leqq \sqrt{\tfrac{1}{2}} + \sqrt{\tfrac{3}{4}} < 2.$$

Also by (5) we see that

$$|(\alpha - u)(\beta - u)|^2 \leqq (x^2 + k^2 + \tfrac{1}{4})^2 - 4k^2x^2.$$

Denoting the right member of this by $F(x, k)$, we prove that $F(x, k) \leqq \tfrac{9}{16}$, and this will complete the proof of the theorem. It suffices to consider $x \geqq 0$, and for $0 \leqq x \leqq 1/\sqrt{2}$ and k fixed

$$F(x, k) \leqq \max \left\{ F(0, k),\ F\left(\frac{1}{\sqrt{2}}, k\right) \right\},$$

because $F(x, k)$ is a quadratic in x^2 whose only critical point is a minimum, and so we need only look at the endpoints $x^2 = 0$ and $x^2 = \tfrac{1}{2}$. Then we compute

$$F(0, k) = (k^2 + \tfrac{1}{4})^2 \leqq (\tfrac{1}{2} + \tfrac{1}{4})^2 = \tfrac{9}{16},$$

$$F\left(\frac{1}{\sqrt{2}}, k\right) = (k^2 + \tfrac{3}{4})^2 - 2k^2$$

$$\leqq \max \{(\tfrac{1}{2} + \tfrac{3}{4})^2 - 2(\tfrac{1}{2}),\ (0 + \tfrac{3}{4})^2 - 2(0)\}$$

$$= \max \{\tfrac{9}{16}, \tfrac{9}{16}\} = \tfrac{9}{16}.$$

5.3. *Minimum values of forms*

With the work of the last two sections as background, we can now write analogues of Theorem 2.3, Corollary 2.4, and Theorem 2.5.

THEOREM 5.4. *Let $a_1x + b_1y + c_1$ and $a_2x + b_2y + c_2$ be linear forms with complex coefficients so that $\Delta = |a_1b_2 - a_2b_1| \neq 0$. Furthermore, suppose that a_1/b_1 is complex irrational and that $a_1x + b_1y + c_1 = 0$ has no solution in complex integers. Then for any given positive ε there are infinitely many pairs of complex integers x, y such that*

$$(7) \quad |a_1x + b_1y + c_1| \cdot |a_2x + b_2y + c_2| < \frac{\Delta}{2} \quad \text{and}$$

$$|a_1x + b_1y + c_1| < \varepsilon.$$

Proof: Generally speaking the proof is directly parallel to the proof of Theorem 2.3, with Theorems 5.3 and 4.3 replacing Lemma 2.2 and Theorem 1.5. However, since the constants are different and there is a slight stricture at one point of the present proof, we will sketch the argument. Equation (5) in the proof of Theorem 2.3 is now replaced by

$$|a_1k + b_1h| < \frac{|b_1|}{|k\sqrt{3}|},$$

and the constant $\frac{1}{2}$ in (7) is replaced by $1/\sqrt{2}$. The requirements (6) of Theorem 2.3 are replaced by

$$\frac{2|b_1|}{\sqrt{3}|k|} < \varepsilon \quad \text{and} \quad \frac{|b_1b_2|}{|k^2|} < \frac{\Delta}{100}.$$

The constant $\frac{1}{4}$ in (10) is replaced by $\frac{3}{4}$, and the argument following (10) now takes the following form. We have

$$\frac{h}{k} = -\frac{a_1}{b_1} + \frac{\delta}{k^2} \quad \text{where } |\delta| < \frac{1}{\sqrt{3}},$$

and hence

$$|a_1k + b_1h| \cdot |a_2k + b_2h|$$

$$= \left|\frac{b_1 \delta}{k}\right| \cdot \left|a_2 k - \frac{a_1 b_2 k}{b_1} + \frac{b_2 \delta}{k}\right|$$

$$= |\delta| \cdot \left|a_2 b_1 - a_1 b_2 + \frac{b_1 b_2 \delta}{k^2}\right|$$

$$\leqq |\delta| \cdot |a_2 b_1 - a_1 b_2| + |\delta| \cdot \left|\frac{b_1 b_2 \delta}{k^2}\right|$$

$$\leqq \frac{\Delta}{\sqrt{3}} + \frac{\Delta}{300}.$$

Hence equation (7) in our present Theorem 5.4 is satisfied because

$$\frac{3}{4}\left(\frac{\Delta}{\sqrt{3}} + \frac{\Delta}{300}\right) < \frac{\Delta}{2}.$$

Finally, equation (11) of Theorem 2.3 has a direct parallel in the present case, with a simple change in the constants.

COROLLARY 5.5. *If θ is complex irrational and α is any complex number such that $\theta x + y + \alpha = 0$ has no solution in integers x, y then for any given ε there are infinitely many pairs of complex integers x, y such that*

$$|x(\theta x + y + \alpha)| < \tfrac{1}{2} \quad \text{and} \quad |\theta x + y + \alpha| < \varepsilon.$$

The proof of this is directly parallel to the proof of Corollary 2.4, with no difficulties whatsoever. Likewise we obtain the following analogue to Theorem 2.5.

THEOREM 5.6. *Let $a_1 x + b_1 y + c_1$ and $a_2 x + b_2 y + c_2$ be linear forms with complex coefficients such that $\Delta = |a_1 b_2 - a_2 b_1| \neq 0$. Then there exists at least one pair of complex integers x, y such that*

$$|a_1 x + b_1 y + c_1| \cdot |a_2 x + b_2 y + c_2| \leqq \Delta/2.$$

As an example where there is only a finite number of pairs of complex integers, consider the forms $x - (1 + i)/2$ and $y - (1 + i)/2$.

5.4. *Further results*

Theorem 5.3 was communicated to the writer by L. C. Eggan and E. A. Maier in response to a suggestion that an analogue to Lemma 2.2

would be useful. Their proof used a covering theorem by Sawyer (1953); the proof given in Section 5.2 uses the simple covering Theorem 5.1 in place of Sawyer's. The writer has extended Theorem 5.1 to the three-dimensional case, that is, the covering of lattice points by rectangular parallelopipeds; this result will be published elsewhere.

Theorem 5.6 is due to Hlawka (1938), who gives precise circumstances for equality. Hlawka's proof, like that given here, is based on the Ford theorem; for proofs avoiding the Ford theorem, see Mahler (1940) and Chalk (1956).

BIBLIOGRAPHY

Th. Bang (1957), On the sequence [nα], n = 1, 2, ···, *Math. Scand.*, **5**, 69–76.

Samuel Beatty (1926), Problem 3173, *Amer. Math. Monthly*, **33**, 159.

J. W. S. Cassels (1954), Über lim $x|\theta x + \alpha - y|$, *Math. Annalen*, **127**, 288–304.

J. W. S. Cassels (1957), An introduction to Diophantine approximation, *Cambridge Tract No.* 45, Cambridge University Press.

J. H. H. Chalk (1955), Rational approximations in the complex plane, *J. London Math. Soc.*, **30**, 327–343.

J. H. H. Chalk (1956), Rational approximations in the complex plane II, *J. London Math. Soc.*, **31**, 216–221.

H. Davenport (1954a), Simultaneous Diophantine approximation, *Proceedings International Congress Mathematicians*, Amsterdam, **3**, pp. 9–12.

H. Davenport (1954b), Simultaneous Diophantine approximations, *Mathematika*, **1**, 51–72.

H. Davenport and K. Roth (1955), Rational approximations to algebraic numbers, *Mathematika*, **2**, 160–167.

L. C. Eggan (1961), On Diophantine approximations, *Trans. Amer. Math. Soc.*, **99**, 102–117.

L. C. Eggan and E. A. Maier (1961), A result in the geometry of numbers, *Mich. Math. J.*, **8**, 161–166.

L. C. Eggan and I. Niven (1961), A remark on one-sided approximation, *Proc. Amer. Math. Soc.*, **12**, 538–540.

L. R. Ford (1925), On the closeness of approach of complex rational fractions to a complex rational number, *Trans. Amer. Math. Soc.*, **27**, 146–154.

G. H. Hardy and E. M. Wright (1960), *An Introduction to the Theory of Numbers*, 4th edition, Oxford University Press, Oxford.

S. Hartman (1949), Sur une condition supplémentaire dans les approximations diophantiques, *Colloq. Math.*, **2**, 48–51.

E. Hlawka (1938), Über die Approximationen von zwei komplexen inhomogenen Linearformen, *Monats. Math.*, **46**, 324–344.

N. Hofreiter (1952), Über die Approximationen von komplexen Zahlen durch Zahlen des Körpers K(i), *Monats. Math.*, **56**, 61–74.

A. Hurwitz (1891), Über die angenäherte Darstellung der Irrationalzahlen durch rationale Bruche, *Math. Annalen*, **39**, 279–284.

A. Khintchine (1935), Neuer Beweis und Verallgemeinerung eines Hurwitzschen Satzes, *Math. Annalen*, **111**, 631–637.

63

J. F. Koksma (1936), Diophantische Approximationen, *Ergebnisse der Mathematik*, Band IV, Heft 4, Springer, Berlin (reprinted by Chelsea).

J. F. Koksma (1951), Sur l'approximation des nombres irrationels sous une condition supplémentaire, *Simon Stevin*, 28, 199–202.

S. Lang (1962), *Diophantine Geometry*, Interscience Publishers.

F. Lettenmeyer (1923), Neuer Beweis des allgemeinen Kroneckerschen Approximationssatzes, *Proc. London Math. Soc.* (2), 21, 306–314.

W. J. LeVeque (1953), On asymmetric approximations, *Michigan Math. Journal*, 2, 1–6.

W. J. LeVeque (1956), *Topics In Number Theory*, vols. I and II, Addison-Wesley.

A. M. Macbeath (1947), The minimum of an indefinite binary quadratic form, *J. London Math. Soc.*, 22, 261–262.

K. Mahler (1940), On the product of two linear complex polynomials in two variables, *J. London Math. Soc.*, 15, 213–236.

K. Mahler (1957), On the fractional parts of the powers of a rational number II, *Mathematika*, 4, 122–124.

K. Mahler (1961), *Lectures on Diophantine Approximation*, Notre Dame, University of Notre Dame Press.

H. Minkowski (1907), *Diophantische Approximationen*, Leipzig (reprinted by Chelsea, 1957).

N. Negoescu (1948), Quelques précisions concernant le théorème de M. B. Segre sur des approximations asymétriques des nombres irrationels par les rationels, *Bull. Ecole Poytech. Jassy*, 3, 3–16.

I. Niven (1956), *Irrational Numbers*, Carus Monograph No. 11, John Wiley and Sons, New York.

I. Niven (1961), Uniform distribution of sequences of integers, *Trans. Amer. Math. Soc.*, 98, 52–61.

I. Niven (1961a), Minkowski's theorem on nonhomogeneous approximation, *Proc. Amer. Math. Soc.*, 12, 992–993.

I. Niven (1962), On asymmetric Diophantine approximations, *Michigan Math. J.*, 9, 121–123.

I. Niven and H. S. Zuckerman (1960), *An Introduction To The Theory Of Numbers*, John Wiley and Sons, New York.

C. D. Olds (1946), Noto on an asymmetric Diophantine approximation, *Bull. Amer. Math. Soc.*, 52, 261–263.

A. Oppenheim (1941), Rational approximations to irrationals, *Bull. Amer. Math. Soc.*, 47, 602–604.

G. Pall (1943), On the product of linear forms, *Amer. Math. Monthly*, 50, 173–175.

O. Perron (1930, 1931), Über die Approximationen einer komplexen Zahl durch Zahlen der Körpers $K(i)$; I, *Math. Annalen*, 103, 533–544; II, *Math. Annalen*, 105, 160–164.

O. Perron (1932), Eine Abschätzung fur die untere Grenze der absoluten Beträge der durch eine reelle oder imaginäre binäre quadratische Form darstellbaren Zahlen, *Math. Zeits.*, **35**, 563–578.

A. V. Prasad (1948), Note on a theorem of Hurwitz, *J. London Math. Soc.*, **23**, 169–171.

D. Ridout (1957), Rational approximations to algebraic numbers, *Mathematika*, **4**, 125–131.

R. M. Robinson (1940), The approximation of irrational numbers by fractions with odd or even terms, *Duke J. Math.*, **7**, 354–359.

R. M. Robinson (1947), Unsymmetric approximation of irrational numbers, *Bull. Amer. Math. Soc.*, **53**, 351–361.

R. M. Robinson (1948), The critical numbers for unsymmetrical approximation, *Bull. Amer. Math. Soc.*, **54**, 693–705.

K. F. Roth (1955), Rational approximations to algebraic numbers, *Mathematika*, **2**, 1–20.

K. F. Roth (1960), Rational approximations to algebraic numbers, *Proceedings International Congress Mathematicians Edinburgh*, 1958, pp. 203–210.

D. B. Sawyer (1953), On the covering of lattice points by convex regions, *Quart. J. Math.*, *N.S.*, **4**, 284–292.

W. T. Scott (1940), Approximation to real irrationals by certain classes of rational fractions, *Bull. Amer. Math. Soc.*, **46**, 124–129.

B. Segre (1945), Lattice points in infinite domains and asymmetric Diophantine approximation, *Duke J. Math.*, **12**, 337–365.

Th. Skolem (1957), On certain distributions of integers in pairs with given differences, *Math. Scand.*, **5**, 57–68.

L. Tornheim (1955a), Asymmetric minima of quadratic forms and asymmetric Diophantine approximation, *Duke J. Math.*, **22**, 287–294.

L. Tornheim (1955b), Approximation to irrationals by classes of rational numbers, *Proc. Amer. Math. Soc.*, **6**, 260–264.

S. Uchiyama (1961), On the uniform distribution of sequences of integers, *Proc. Japan Acad.*, **37**, 605–609.

Index